The Audubon Wildlife Treasury

The Audubon
Wildlife Treasury

Edited by LES LINE

Drawings by Chuck Ripper

Published in cooperation with the National Audubon Society
by J. B. LIPPINCOTT COMPANY · Philadelphia and New York

All of the material in this book originally appeared in somewhat different form in *Audubon* magazine. Copyright © 1967, 1968, 1969, 1970, 1971, 1973, 1974, 1975 by the National Audubon Society.

"Journey Among Barbarians" is reprinted by permission of Harold Ober Associates Incorporated. Copyright © 1967 by National Audubon Society.

"Interview with a Shrew" is reprinted by permission of Harold Matson Co. Inc. Copyright © 1975 by Jack Schaefer.

"Tamiasciurus—the Harvester" is reprinted by permission of Collins-Knowlton-Wing. Copyright © 1968 by Hal Borland.

Grateful acknowledgment is made to the following publishers and authors for permission to use and adapt material which subsequently appeared in book form:

"Windows on Wildlife" is reprinted by permission of Dodd, Mead & Company, Inc., from *A Naturalist Buys an Old Farm* by Edwin Way Teale. Copyright © 1972 by Edwin Way Teale.

"Immortality" is reprinted by permission of Simon & Schuster, Inc., from *Mosquito Safari* by C. Brooke Worth. Copyright © 1971 by C. Brooke Worth.

"Coming of the Giant Wasps" is reprinted by permission of Charles Scribner's Sons from *All the Strange Hours* by Loren Eiseley. Copyright © 1975 by Loren Eiseley.

U.S. Library of Congress Cataloging in Publication Data

Main entry under title:

The Audubon wildlife treasury.

  CONTENTS: Krutch, J. W. Unnatural history.—Ford, C. Journey among barbarians.—Matthiessen, P. In the dragon islands. [etc.]
  1. Zoology—Addresses, essays, lectures.
2. Animals, Legends and stories of. I. Line, Les.
II. Audubon magazine.
QL81.A88   591   76–10708
ISBN–0–397–01149–0

In memory of Carl Hermann,
stalwart protector of Michigan's wildlife,
whose path crossed mine in the formative years.

CONTENTS

Audubon's Photographers
Portfolio One follows page 64
Portfolio Two follows page 128

FOREWORD

A decade ago, in an essay for *Audubon* that is even more pertinent today, the great naturalist-philosopher Joseph Wood Krutch expressed dismay at what he viewed as "The Demise of Natural History." What particularly distressed him was the definition of natural history in *Webster's Third New International Dictionary* as "a former branch of knowledge embracing the study, description, and classification of natural objects." While he considered that obituary to be somewhat premature, "the tendency of much official science has been in a direction which makes it approximately true," he wrote. "And that means nothing short of a calamity to those of us whose attitude toward nature is both esthetic and emotional as well as scientific, and to whom, for that reason, conservation is a primary concern."

The love of nature, Dr. Krutch continued, "provides the most effective motive for the preservation of our wild heritage." But in this century, the development of biology "has led it further and further away from the methods and concerns of natural history. It has made the biologist less and less a man of the out-of-doors, more and more a man of the laboratories. And in the laboratory he has been led further and further away from everything which tends to establish an empathy between himself and the subjects he studies. He has emphasized purely mechanistic interpretations of animal behavior and has tended to transform biology itself into biochemistry. He has retreated, not only from the woods and fields, but from everything that is, or can be, part of the everyday experience. Instead of watching a bird in the forest or on the seashore, the modern biologist is more likely to be found peering inside a cell or trying to analyze the chemical constitution of the chromosome and its genes."

As an example of the demise of natural history, Dr. Krutch recalled asking the botany professor of a small college about a flowering tree he had seen on campus. The professor replied in quite condescending terms that he was a cytologist—a specialist in the study of cells—and could not recognize a dozen plants by name. That kind of attitude toward nature is harmful mainly to the professor's students and to the generations they may influence. But such is not always the case. Two years ago a noted American ornithologist ran afoul of federal laws for encouraging collectors around the world to violate bird-protection statutes in order to provide him with eggs of rare species for a research project. Upon paying a substantial fine, he was quoted as saying his research was "more important than the growth into maturity of a peregrine falcon," and that "bird-conservation groups are composed of people ruled by their emotions and with little or no knowledge of bird populations."

How wrong the ornithologist was. As Dr. Krutch commented, "One need not be ignorant of science in order to love nature. Some of us would, nevertheless, rather be ignorant of biochemistry than as incapable of loving nature as some biologists apparently are." For eloquent support, he turned to the writings of William Morton Wheeler, the renowned American entomologist of the early twentieth century and a specialist whose concern for natural history was as great as his technical expertise.

"Why animals and plants are as they are," Wheeler said, "we shall never know; how they came to be what they are, our knowledge will always be extremely fragmentary; but that organisms *are* as they *are*, that apart from members of our own species they are our only companions in an infinite and unsympathetic waste of electrons, planets, nebulae, and stars, is a perennial joy and consolation."

As I said at the beginning, Joseph Wood Krutch's essay is even more pertinent today than when written ten years ago. There are two reasons. First, there is no evidence of any exodus, among biologists, from their laboratories and test tubes to the woods and fields. Indeed, the journals of the various biological sciences are chockablock with papers on esoteric research projects by technicians who may never have seen their study subjects living and free, and couldn't care less. Second, conservationists have been forced, in these difficult times, to abandon the love of nature as a primary justification for saving wild places and wild things. Because economic gains are constantly being weighed against environmental losses, and because businessmen and bureaucrats and politicians always have their thumbs on the scales, it is no longer enough that a salt marsh, for instance, be saved for its esthetic values.

Instead, one must assess how many dollars each acre of marsh is worth for commercial and sports fishing, for aquaculture, for tertiary waste treatment, for storm protection, for production of oxygen, and judge these figures against the value of the land for oil refineries or vacation homes. That is all well and good, for any facts that might sway decisions toward the side of preservation are welcome. But at the same time, conservationists are being urged by some of their leaders to forever curb emotional arguments in behalf of unspoiled nature. Thus I fear that one day soon the environmental movement, like the biological sciences, will be peopled by solemn technicians who view the natural world with calculators and computers, who also hold mere nature lovers in contempt, and who have never experienced the "joy and consolation" that is the historic stanchion of their cause. And that too will be a calamity.

In the meantime, I must report that the demise of natural history has not *yet* occurred. There is still, among both biologists and laymen, a small but hard core of men and women who do not shrink from being called naturalists, whose works and writings reflect both the esthetic and the scientific—men and women who would be more comfortable in the company of a Joseph Wood Krutch or a William Morton Wheeler than in the presence of modern biochemists. Many of them are represented in this collection of twenty articles drawn from the pages of *Audubon* during the eleven years of my editorship. I can only pray that the time will never come when such volumes, left dusty and unopened on library shelves, are the only remaining evidence of "a former branch of knowledge" embracing the study—and love—of nature.

LES LINE
Editor, AUDUBON

UNNATURAL HISTORY, BY JOSEPH WOOD KRUTCH *Why people of long ago believed that only a virgin could capture a unicorn, that salamanders lived in fire, that amorous mermaids sang to woo sailors, and that antlions must certainly go forever hungry.*

Natural History, in the modern sense of the word, is a very recent thing. But Unnatural History goes back to the dawn of civilization. For more than two thousand years—from 500 B.C. to the seventeenth century—most of the things which the learned as well as the unlearned knew about animals weren't so, though they were often quite interesting in a believe-it-or-not way and very useful to moralists as well as to poets.

Even today, rustics believe in the hoop snake whose picture, tail in mouth, appears in medieval bestiaries. We continue to use metaphors or similes we no longer understand—as when we say "lick into shape" without knowing that for centuries everybody *knew* that bear cubs are shapeless until their mother molds them with her tongue. Books of unnatural history were for more than a thousand years favorite reading matter for those who could read, while the illiterate learned the same strange pseudofacts from the sculptured ornaments of Romanesque churches where unnatural history was a subject exploited almost as often as stories from the Bible.

No doubt some of the misinformation survived because it was so interesting that everybody wanted to believe it. Something of the same sort doubtless accounts for the fact that, for at least fifteen hundred years, the unicorn, the mermaid, and other wonderful but nonexistent animals were as solidly established in what passed for zoology as, for instance, the rhinoceros and the hippopotamus. Neither of the latter had been seen by Europeans except on very rare occasions since the days of the Roman Empire, when they had been brought from Africa to be slaughtered in the arena. In appearance both were more improbable and a good deal less attractive than the unicorn.

I do not suppose that anyone today would maintain that the unicorn is a real animal. In fact, one zoologist has proved that no mammal could possibly have a horn in the middle of its forehead because of the bone suture located there. As a consequence, the rhinoceros, which does insist upon having one horn only, has to place it at the end of its nose. But we still wish we could believe in the unicorn and for that matter in the mermaid; and because pictures of both appear in one form or another as regularly as any animal except the dog and the cat, we recognize them as readily as we do an elephant or a crocodile. If we are inclined to look down upon our ancestors for believing in the unicorn and the mermaid without having seen either, we should remember that anyone who would believe in a giraffe on any evidence less convincing than an ocular demonstration would believe anything!

If such undesirable creatures as the basilisk, the manticore, and the werewolf continued to haunt the pages of the old naturalists almost as persistently as the unicorn, the explanation may be that we also enjoy making our flesh creep. The basilisk who killed with a look and the man-eating manticore whose double row of teeth were designed like the teeth of Little Red Riding Hood's wolf—"the better to eat you with, my dear"—were terrifying enough. As for the werewolf, there is another aspect. Both the early Greeks and the Navajo Indians on their Arizona reservation agree not only that the werewolf is a reality, but also on an interesting detail: anyone who eats human flesh will presently find himself in wolf's clothing. This raises an interesting question for debate between diffusionists and the champions of independent cultural invention.

A single erroneous belief, like that concerning the unlicked cub, illustrates some of the ways in which such errors arose, persisted, and flourished. Sometimes the unnatural history of a real or imaginary animal began with a simple misinterpretation of an observed act, as in the case of the unlicked bear cubs. Actually, they are born in caves during the quasihibernation of the female and they are in a sense premature, or at least so disproportionately small that a 200-pound bear may give birth to a cub weighing as little as six ounces. The belief that these tiny creatures had to be licked into shape might easily follow from a casual observation, and it was already firmly established at the time of Pliny the Elder (First Century A.D), from whom it was copied by nearly every writer on animals during the next fifteen hundred years.

The equally well-known belief that the salamander lives in fire arose from a similar misinterpretation. Benvenuto Cellini's eyewitness account of this remarkable creature frisking midst the burning logs of the fireplace was honestly intended, though there is no doubt that what he really saw was the death agony of a moisture-loving salamander which had been tossed into the flames with the damp logs among which it had been living. Some centuries before this, Marco Polo had brought back confirmatory evidence from the Orient in the form of specimens of incombustible salamander's wool or, in other words, bits of asbestos.

Sometimes, on the other hand, it was a word rather than a fact which was misinterpreted. When a medieval scholar came across the word "antlion," which occurs in the Septuagint, he tried to imagine what sort of creature it might refer to and came up with the notion of an odd hybrid, half-ant and half-lion. Moreover, in at least one instance he drew a picture of what he thought it might look like and also concluded that it must usually die of starvation since there was no food suitable for both an ant and a lion—a fact which put the unfortunate creature in the same situation as that of Alice's bread-and-butterfly.

Though the mermaid and the unicorn are today the most vividly alive of all mythical creatures, their origins and their histories are very different. There is nothing obviously improbable about a horselike creature with one horn, whereas a half-fish half-woman immediately rouses in the rational mind puzzling questions of anatomy and physiology—"too much fish to eat and too little woman to love," as modern cynicism has it. This difference is enough to suggest what is actually the case, namely that the unicorn begins with verbal confusions and mistaken observation while the origin of the mermaid is myth.

Some translations of the Old Testament refer to something called a unicorn, and some ancient travelers casually refer to some one-horned creature. In the first case the reference was probably to the rhinoceros, the other cases merely the result of mistaken observations. In any event it was not until the Dark Ages or the Middle Ages that these vague allusions did much more than create the assumption that a little-known and not especially interesting animal did exist somewhere. Then superstition and perhaps deliberate charlatanry got to work. The unicorn was said to be an extremely rare animal whose horn had magical properties—as an antidote to poison and as an aphrodisiac. Powdered unicorn horn (rhinoceros, perhaps) was sold by druggists, and in a number of churches unicorn horns were kept as great treasures. In many cases these were the monstrously overgrown teeth of the narwhal which are twisted in exactly the fashion of the horn on the familiar representations of the unicorn.

With this tangible evidence of the creature's existence, the medieval mind set to work inventing for the unicorn a rather varied natural history—including the fact that it fought with elephants and stuck them in the belly with its one horn. Much more interesting was the notion, repeated in many bestiaries, that the unicorn could be captured in only one way: If a virgin be planted in the woods, the unicorn will lay its head in her lap and allow itself to be taken. This is the story most often told in words and pictures, notably in the series of tapestries (woven about 1500) now exhibited at The Cloisters in New York City. By this time the story had become more romance than natural history and is supposed somehow to symbolize the marriage of the unicorn, Charles VIII of France, with the virgin, Anne of Brittany.

The history of the mermaid is even more complex. Images of a Phoenician water god and goddess look exactly like our mermaids, but there is no evidence that the stories we know were connected with them. On the other hand, the sirens who lured the sailors of Odysseus are obviously related, at least, to the mermaid of the medieval legend. But Homer's sirens are half-bird half-woman, not half-fish.

What seems to have happened is this: Sometime during the centuries just preceding the Christian era, the form of the Phoenician water goddess was attributed to creatures who behaved like sirens and who began to appear as such in Hellenistic and Roman sculpture. In this shape, mermaids appear also in scores of Romanesque churches where they sometimes have both wings and fish tails. In the churches they seem to be sirens by nature, luring sailors to destruction and symbolizing the lusts of the flesh.

Then a contrary romantic arose, and the conflict between the two is still evident in many of the stories which continued to be invented or retold from medieval times to the present. In their most romantic Christian form they turn upon the fact that mermaids sing alluringly to sailors, not in order to destroy them but to win them as husbands—since it is only by marrying a human being that these only half-human creatures can acquire immortal souls. And it is the amorous rather than the destructive mermaid (or, sometimes, merman) who appears most frequently in modern literature as in Hans Christian Andersen's *The Little Mermaid* and Matthew Arnold's "The Forsaken Merman."

Some legends survived because it was not easy to check them and because they had the authority of the ancient writers to back them up, but the most extravagant reached their full development in the medieval bestiaries where their wild elaboration was due to something quite different from either ignorance, romantic fantasy, or simple love of the marvelous. It was the consequence of a theory of the universe so different from ours or that of classical antiquity that we can hardly understand it.

Among the Greeks who wrote about animals, Herodotus had a strong sense of fact and Aristotle was a scientist in the modern meaning of the term. Among the Latins, Pliny, perhaps the most influential writer about natural history who ever lived, was uncritically credulous and passed on hundreds of myths. But he was not mystical and, like Aristotle, his concern was with what he believed to be objective fact.

The Middle Ages, on the other hand, took it for granted that everything in the universe had been created to serve man, either directly or as a moral lesson, and writers always looked not for objective facts but for the intention everything was supposed to embody. This meant, for instance, what ailment a plant would cure or what a supposed fact of natural history could teach. The monk in his cloister did not doubt any alleged fact if a moral could be drawn from it. Aristotle had recorded the report that an elephant has no knees. The medieval mind seized upon this often-repeated statement and expanded it into an elaborate allegory illustrative of the fact that fallen man could not rise by his own efforts but must be lifted up by divine power. It seemed also that the common belief that lion cubs were born dead and were brought to life three days later, when the male breathed upon them, must be true because it was an obvious allegory of Christ's resurrection.

In the same way, the ancient story that a beaver pursued by hunters for the sake of his odoriferous testicles (actually scent glands) would bite them off in order to escape further pursuit was intended to teach simple man that, as one writer put it, "every man who inclines toward the commandment of God . . . must cut off from himself all vices . . . and cast them from him in the Devil's face." Even the tale of the solitary phoenix, reborn from time to time in flames, was accepted as an allegory of Christ's resurrection, and by one somewhat later writer rejected for an equally unlikely reason. God commanded all creatures to increase and multiply. He would certainly never have created a bird of which no more than one ever existed.

Edward Topsell's *The History of Four-footed Beasts* (first published in 1607 and based largely on the earlier original by the Swiss naturalist Gesner) is often called the first book in English which can be described as serious zoology. Such it actually was in intention, but it is largely a compilation from older writers; it repeats most of the old tales, sometimes rejecting them but often accepting them, and it includes a long article on the unicorn as well as shorter accounts of several other mythical creatures. Indeed, it actually adds to the standard medieval zoo a horrible creature from the New World called the su which, from the drawing supplied, appears to be a nightmare version of the innocent opossum. Nevertheless, it did drop most of the medieval allegories.

If you follow the rise, development, and gradual decay of unnatural history, the course is always the same: first there is an alleged fact, often of Oriental origin, and usually vouched for by some ancient Greek or Roman authority. During the Dark Ages and Middle Ages, this alleged fact is elaborated in the interest of a moral lesson which continues to be repeated in book after book until the Renaissance, when the moral is usually dropped and only the classical fable repeated. Finally, in the seventeenth century, the fable also is rejected in its entirety either by an appeal to common sense or, sometimes, as the result of an actual investigation. Before the eighteenth century was out, natural history founded either directly or ultimately on fresh observations was firmly established in such different forms as Buffon's learned works and the amateur observations in Gilbert White's *Natural History and Antiquities of Selborne.*

That does not, however, mean that there is not a good deal of unnatural history still about—as, for example, in the Down Easter's firm conviction that a coon cat is half-raccoon. Nor, for that matter, have we ceased to create wonderful-animals-that-never-were—as witness the Loch Ness monster and the abominable snowman, not to mention the Martians who hover over us in flying saucers.

JOURNEY AMONG BARBARIANS, BY COREY FORD

Countless visitors to the parks and forests of the American West have been awed by the large, brilliantly colored, and audacious bird known as Steller's jay. But few among them have ever heard of the young German naturalist for whom the species is named, and fewer still would recall his great scientific discoveries on the first Russian voyage to Alaska.

*Who has earned a greater or more precious glory for his name
than he who undertakes journeys among the barbarians?*

LINNAEUS

"I have fallen in love with nature," Georg Wilhelm Steller wrote in his journal in 1741. It was this total dedication to his calling which led him to endure incredible hardships and cold and shipwreck during Bering's voyage of discovery to Alaska. He was only thirty-seven when he died; but his accomplishments, in the brief span of his lonely life, will make him live forever.

He was the first white man to sight the mainland of Alaska, the first ever to set foot on the soil of northwest America. He was Alaska's pioneer naturalist, discoverer of numerous species which bear his name today: Steller's sea lion, Steller's jay, Steller's eider, Steller's sea eagle, the legendary Steller's white sea raven. He found and recorded Steller's greenling, our brilliantly colored rock trout. He wrote the only description of the giant northern manatee, called Steller's sea cow, which became extinct shortly thereafter. He reported an even stranger creature never again seen by man—Steller's sea monkey—which lives only in this young German's exact field notes.

Steller's journal reveals his complex and contradictory nature: hypersensitive himself and yet insensitive to the feelings of others, indefatigable and brilliant but dogmatic and without tact, an irascible genius who lacked the saving grace of humility and who was unable to tolerate any difference of opinion.

His driving ambition led him from his native Bavaria, where he had studied botany and the closely allied subject of medicine, to St. Petersburg and thence across Siberia and Kamchatka—a third of the distance around the globe —to join the crew of the *St. Peter* and sail on commander Vitus Bering's epic journey to *Bolshaya Zemlya,* the Great Land, Alaska.

He had a scathing contempt for his Russian shipmates, whom he considered little more than barbarians; but when scurvy broke out on the voyage, his training as a physician enabled him to locate antiscorbutic herbs which saved their lives, and he tended the survivors devotedly after their shipwreck on Bering Island.

Commander Bering was already broken in health when he sailed. Ten heartbreaking years of preparation for the expedition had robbed him of the zeal for adventure, even the will to live. When the *St. Peter* made landfall on July 15, 1741, in the vicinity of what is now Cape St. Elias, he merely shrugged at the sight of the New World. His sole thought was to take on fresh water and sail back to Kamchatka before the fall storms closed in. Only after passionate pleading was young Steller permitted to go ashore for half a day on the present Kayak Island, to observe the flora and fauna of the unknown continent.

"The only reason why we did not land on the mainland is a sluggish obstinacy, a dull fear of being attacked by a handful of still more timid savages, and a cowardly homesickness," Steller noted bitterly in his journal. "Ten years the preparation for this great undertaking lasted, and ten hours were devoted to the work itself."

Perhaps no other naturalist in history ever accomplished so monumental a task under such difficulties and in so little time. With his Cossack hunter Thoma Lepekhin, Steller combed the island all the day, gathering botanical specimens. It was four in the afternoon when he spread his findings around him on the sandy beach and began to enter in his notebook the results of his explorations. When the yawl came to pick him up at five o'clock, he had completed his exhaustive report, the first scientific paper ever written on Alaskan natural history. The original manuscript still exists in the archives of the Academy of Sciences, U.S.S.R., and a reproduction of the classic document is preserved in the Library of Congress in this country.

Some plants in his collection were already familiar to him from his earlier investigations in Kamchatka. He identified the upland cranberry, the red and black whortleberry, and a shrub which he called the scurvy berry, probably the black crowberry. He was more enthusiastic over "a new and elsewhere unknown species of raspberry," the salmonberry of Alaska and the Pacific Northwest.

He had ordered his Cossack companion to shoot some of the native birds he had noticed, easily distinguished from the European and Siberian species by their particularly bright coloring; and on his return Lepekhin "placed in my hands a specimen," he wrote, "of which I remember to have seen a likeness painted in lively colors and described in the newest account of the birds of the Carolinas."

Steller's incredible memory had recalled a handcolored plate of the eastern American blue jay in Mark Catesby's *Natural History of Carolina, Florida and the Bahama Islands,* which he had seen years before back in Germany; and he identified Lepekhin's find as its West Coast cousin, known today as *Cyanocitta stelleri,* or Steller's jay. Now his last doubts about the land they had discovered were resolved. "This bird proved to me that we were really in America."

Bering's premonition of disaster turned out to be only too accurate. Autumn storms closed in, awesome williwaws tore the canvas to shreds, giant waves battered the *St. Peter* until the crew expected that their vessel would founder at any moment. For four months they struggled blindly through the fog and rain of the North Pacific. Scurvy, the bane of eighteenth-century mariners, struck the weary crew, deaths became almost a daily occurrence, and by November only three or four sailors were able to drag themselves from their bunks to man the ship. They had lost all sense of direction. When an island appeared on the horizon, they were confident it was the mainland of Kamchatka and made for it eagerly. A heavy surf scraped the *St. Peter* over a submerged reef, and the wrecked ship came to rest in a partly protected bay, never to sail again.

Steller was certain they had not reached the mainland. His conviction was strengthened when he rowed ashore in the longboat.

"We had not yet reached the beach when a strange and disquieting sight greeted us, as from the land a number of sea otter came toward us, which from a distance some took for bears, others for wolverines." The fearlessness of the otters was proof to him that they had never seen man before, and therefore this was an uninhabited island.

The sea otters, so tame they could be slain with clubs, were virtually the only source of food for the castaways in the first months. The flesh was rank-tasting and tough and sinewy, difficult for the scurvy victims to chew with their swollen gums and loosened teeth. The disease was taking a terrible toll; more than a third of the crew had succumbed, and Commander Bering died shortly before Christmas; he was buried on the island which bears his name today.

Gradually, however, the fresh water and meat had its effect on the others; and antiscorbutic roots and bulbs which Steller dug from the frozen earth helped to check the epidemic. While the survivors were hunting more otters, Steller made careful field notes about this marine mammal's behavior and characteristics, the first and still the best description of the sea otter.

"Altogether it is a beautiful and pleasing animal," he wrote, "cunning and amusing in its habits, and at the same time ingratiating and amorous. They prefer to lie together in families, the male with its mate, the half-grown young and the very young sucklings all together. The male caresses the female by stroking her, using the forefeet as hands; she, however, often pushes him away from her for fun and in simulated coyness, as it were, and plays with her offspring like the fondest mother.

"Their love for their young is so intense that they expose themselves to the most manifest danger of death. When their young are taken away from them, they cry bitterly, like a small child, and grieve so much that, as we have observed from rather authentic cases, after ten to fourteen days they grow as lean as a skeleton, become sick and feeble, and will not leave the shore."

The long cruel winter ended at last, and spring arrived with a rush of wings. Suddenly the April sky was filled with myriad migratory birds, returning to their northern nesting grounds on Bering Island. Murres and kittiwakes stirred the slumbering shores to life. The silent hills resounded to the harsh laughter of red-throated loons, and the low liquid notes of the dainty snow bunting were a familiar melody which carried Steller back to his boyhood days in Bavaria.

The drifts were still deep in the valleys when the first red-footed black guillemots came in from the sea and settled on the rocky ledges of their old rookeries. White-breasted auks followed them, whistling shrilly from the cliffs behind the camp. Here and there, where the snow had melted on the slopes, Lapland longspurs chirped their pessimistic refrain, as though brooding over the past days of gloom and fog, and were answered by the wildly joyous song of skylarks heralding the end of winter.

Day and night the V-shaped skeins of wild geese wove their cobwebby patterns across the sky, baying like packs of hounds as they swept overhead. Ducks in countless numbers—mallards and pintails and shovelers and teal—stooled in to join the graceful harlequins which remained on the island year-round.

Immense rafts of bright eiders rocked on the ocean, half a mile from shore; they too had wintered here, but soon would head farther north to breed in the highest Arctic latitudes. Their color was generally white, Steller observed, with brown underparts, as though they had skidded through the mud. They were extremely shy and flew at terrific speed, outdistancing any other duck, and he was unable to collect a specimen. Later Peter Simon Pallas, the noted German naturalist, named the species Steller's eider, *Anas stelleri,* in honor of the first naturalist to record and describe it.

As the snow receded, palatable herbs sent up their first tender shoots, and under Steller's direction the men dug medicinal plants from the tundra: Kamchatkan sweet grass, the root of which resembles the parsnip and is as edible as the stalk; the delectable bulb of the Sarana lily; sea lungwort, brooklime, and bitter cress—all of which gave strength to their scurvy-racked bodies. Their spirits rose, and with renewed courage they set about the herculean task of dismantling the wreck of the *St. Peter* and building a smaller vessel to carry them back to Kamchatka.

During May and June, while the beach before their camp echoed to the busy sound of saws and hammers, Steller spent his time exploring the island more extensively, compiling field notes on the local plants and birds. He had made himself a stout pair of boots out of sealskin, with the fur turned inside, and a coat of the black skins of newborn fur seals. He carried his writing paper and quill pen and inkpot in a waterproof pouch fashioned of seal intestines. With the Cossack Lepekhin, he wandered over the alpine meadows bright with wild roses and violets and golden-flowered rhododendron. Anemone and purple fireweed dotted the lower slopes of the hills, and the wet sphagnum bogs were white with the bloom of cloudberries. Altogether he listed two hundred eleven varieties of plants, most of which were also to be found in eastern Siberia and in the mountain regions of Europe.

The birds on Bering Island were likewise similar to those of Kamchatka, with several notable exceptions. One was a "white sea raven . . . impossible to reach because it only alights singly on the cliffs facing the sea"; it has never been identified or seen again. Another was "a special sea eagle with white [fore]head and tail. In the highest rocks overhanging the sea, it constructs a nest of two ells [about five feet] in diameter, composed of twigs gathered from a great distance, and strewed with grass in the center, in which are one or two eggs, in form, magnitude, and whiteness very like those of a swan. At the beginning of June they have young ones that are completely covered with white down."

His discovery is known today as Steller's sea eagle, one of the four American eagles, so rare in Alaska that only a few have been recorded there since Steller's time.

Another he described as "a special kind of large sea raven with a callow white ring around his eyes and red skin about the beak, which is never seen in Kamchatka, and occurs only on the rocks near Steller's Cave." The spectacled cormorant, one of Steller's most sensational finds, was flightless like a penguin, its stubby wings too small to lift its heavy body, and so helpless that within a hundred years it was totally exterminated by hunters. This ungainly survivor of prehistoric times was as large as a goose, weighing up to fourteen pounds, "so that one single bird was sufficient for three starving men."

Steller prepared it by encasing the bird, feathers and all, in a mold of clay, and baking it in a heated pit to make it tender. Despite the fact spectacled cormorants were extremely plentiful—Steller called them "copiosissimi"—there are only six mounted specimens in existence, and museums prize it more highly than the great auk.

On the side of the island facing Kamchatka, countless herds of fur seals and sea lions "covered the whole beach to such an extent that it was not possible to pass without danger to life and limb." The great rookeries had never been disturbed by man, and Steller had a unique opportunity to study the animals in their natural state. He chose a slight elevation in the center of the rookery and built a flimsy blind of driftwood. Here he remained concealed for six consecutive days, surrounded on all sides by an undulating gray ocean of massive bodies, slumbering or fighting or scratching themselves lazily with the long flexible fingers on their hind feet.

"If I were asked to state how many I have seen on Bering Island," he estimated conservatively, "I can say without lying that it is impossible to make any computation. They are innumerable." His ears rang with the deafening chorus of babies bleating like lambs, the mothers whinnying in a higher register, here and there a ferocious old bull—called *sikatch* or beach master—bellowing his deep bass challenge to anyone who might dare raid his harem.

Crouched in his blind, his writing pad balanced on a knee, Steller composed on the spot the definitive chapter on fur seals, which covers fourteen pages of his famous monograph *De Bestiis Marinis,* published by the St. Petersburg Academy ten years after his death. He even dissected and analyzed the outer and inner structure of a large bull, his detailed measurements and anatomical observations occupying another thirteen pages. So thorough and accurate was his account, written under impossibly difficult conditions, that when the United States took possession of the Pribilof Islands a century later, the government experts who surveyed the fur seal herds could make no corrections or changes in Steller's classic description.

Ever since the stranding of the *St. Peter,* the hungry survivors had cast longing eyes at an abundant supply of fresh food which drifted tantalizingly under their very noses, just offshore. These strange monsters, which the Russians called sea cows, came close to the beach with each flood tide to crop on the pastures of seaweed. Using all their strength and ingenuity, the sailors managed to harpoon one of the creatures and drag it on land, beating it to death with clubs and hacking it apart with bayonets and knives.

The flesh proved to be of superlative quality. The fat, which covered the body to a thickness of four inches, was "glandular, firm, and shiny white, but when exposed to the sun takes on a yellowish tinge like May butter. Both the smell and the taste of it are delicious, and it is beyond comparison with the fat of any marine animal, and even greatly preferable to the meat of any quadruped, since in addition it will keep ever so long even during the hottest days without becoming rancid or strong-smelling. Melted, it tastes so sweet and delicious that we lost all desire for butter. In taste it comes pretty close to the oil of sweet almonds. . . . The meat, when cooked, although it must boil rather long, is exceedingly savory and cannot be distinguished easily from beef. The fat of the calves is so much like fresh lard that it is hard to tell them apart, but their meat differs in no wise from veal."

More important to Steller than the new source of food was the opportunity to examine more closely this fabulous creature. "To the navel it resembles a land animal," he noted. "From there on to the tail, a fish." The corrugated outer skin was blackish brown, full of grooves and wrinkles, and the head suggested "in some measure a buffalo head, particularly as concerns the lips." Having no teeth, it masticated its food like a bovine with the horny plates of its gums.

Steller's description of this long-extinct creature, more than any other achievement, has made his name immortal. All that science knows of it today is the chapter in *De Bestiis Marinis,* written on Bering Island more than two hundred years ago by the only naturalist who ever saw the northern manatee alive.

Steller was determined to complete his description by analyzing the sea cow's inner structure and anatomy; and on July 12, a climactic moment in his career, he set about dissecting a mammoth carcass on the beach. According to his estimate, it weighed about 8,000 pounds; the heart alone tipped the scales at 36¼ pounds. The stomach was "of amazing size, 6 feet long, 5 feet wide, and so stuffed with food and seaweed that four strong men with a rope attached could scarcely move it from its place and drag it out." Despite the difficulties, he did not resort to field notes, but wrote down in concise Latin every anatomical detail, precisely as it was to appear in his monograph:

"The eyes of this animal in spite of its size are not larger than sheeps' eyes, [and are] without eyelids. The ears are so small and hidden that they cannot at all be found until the skin has been taken off, when its polished blackness reveals the ear opening, hardly enough for the insertion of a pea. . . . The feet consist of two joints, the extreme end of which has a rather close resemblance to a horse's hoof, and I hesitate whether to call them hands or feet for the reason that with the exception of the birds we do not have a single two-footed animal.

"With these front feet, on which neither fingers nor nails can be distinguished, the animal swims ahead, knocks the seaweed from the rocks on the bottom, and, when lying on its back getting ready for mating, one embraces the other as with arms. Under these forefeet are found the breasts, with black, wrinkled, two-inch-long teats, at the end of which innumerable milk ducts open. When pulled hard these ducts give off a great amount of milk, which surpasses the milk of land animals in sweetness and richness. . . . The belly is roundish and very distended and at all times stuffed so full that at the slightest wound the entrails at once protrude with much whistling. . . . From the genitals on, the body suddenly decreases in circumference. The tail itself becomes gradually thinner toward the flipper, which serves as hind feet [and] is horizontal as in the whale and the porpoise."

He added with keen perception: "From the head of their maniti the Spaniards are said to take out a stone-hard bone. This I have vainly searched for in so many animals that I have come to think our sea cow may be a different kind."

At infinite pains, Steller prepared a specimen of a young sea cow, consisting of a skin stuffed with dry grass and a complete skeleton—an item which would be beyond price today—and similar mounts of a sea otter and fur seal and sea lion. But space was limited on the makeshift vessel, and he was forced, despite his protests, to leave everything behind when they sailed. All he salvaged was a pair of the horny masticatory plates which served the sea cow instead of teeth. These palatal plates are now in the museum of the Academy of Sciences at Leningrad, the only ones in the world.

The tiny vessel reached Kamchatka on August 27, 1742, more than sixteen months after Bering had departed on his tragic voyage. With the common danger over, Steller's dislike of his barbaric Russian shipmates was renewed, and he left the crew and set out on foot across Kamchatka. For several years he wandered alone in northern Siberia, disillusioned and bitter, convinced that his work as a naturalist had come to nothing. The manuscripts he had forwarded to St. Petersburg had not even been acknowledged. The only result of the expedition had been to open the Aleutians to ruthless fur hunters who would plunder the islands until the herds of sea otters were obliterated. In less than a quarter of a century, the last sea cows on Bering Island were slaughtered and the species exterminated for all time.

Like Bering, Steller had lost the will to live. He succumbed to pneumonia on November 12, 1746, in the north Siberia town of Tyumen, and was buried in a shallow trench on a bluff above the Tura River, with a large stone on the grave to protect his remains from scavenging dogs and wolves.

Twenty-four years later Peter Simon Pallas visited Tyumen and made a pilgrimage to the bleak burial site on the bluff. Atop it was the stone "which will be seen," Pallas wrote, "until the Tura has eaten away the high bank on the spot on which it stands, when Steller's bones will be mingled with the mammoth bones on its farther shore."

There is no trace of the grave today.

IN THE DRAGON ISLANDS, BY PETER MATTHIESSEN

On the schooner Golden Cachalot, *launched in England only a few years after Darwin's death, a naturalist-poet sails from island to island in the legendary Galápagos archipelago, diving with marine iguanas, exploring harsh lava beaches, climbing to the volcanic domain of the giant tortoise, recording in his journal his impressions of this Pacific birthplace of the theory of evolution.*

Seen from the air as the migrant plover sees it, the Galápagos archipelago, six hundred miles west of the coast of Ecuador, appears as a world apart. In this cool season of *garua*—it is late October now—when the arm of the cold Humboldt Current called the South Equatorial Current flows westward through the islands, the greater volcanoes are shrouded in mist that hides the green highlands from view, and elsewhere, isolated cones and rock hulks drift away from one another on a silent silver sea without a sail. The burnt shores might be rising from the methane mists of the Precambrian, before the Earth had life. To these volcanic rocks that rose boiling from submarine plateaus perhaps a million years ago, to burst in white roil from the blue Pacific, winds brought the first air-borne spores of fern and lichen, the aeolian small spiders, and the tiny land snail, *Tornatellides,* whose nearest relatives are two thousand miles away. Later more seeds and insects must have come on the feet of the first seabirds to sight the raw new cones on the horizon.

On Baltra, off the northern shore of Santa Cruz, sits one of five small settlements scattered here and there in the south islands (the five islands north of the equator are all small and without water). In World War II, as a defense for the Panama Canal, the United States bulldozed Baltra flat to make an air base, and the skeletal remains of war—rust, broken concrete, derelict machinery, staring gun emplacements—may never be camouflaged, since plants are spare and scattered here, as is all other life in these sea deserts.

Much of a few hours' walk on Baltra was spent with a flock of ground finches, *Geospiza* of three species, all very similar except in size and proportions of the bill, and comfortable in their mixed flock, as if the distinctions that we make between them were of less significance to finches than to man. In addition to courtship and nesting habits, all share the short tail and weak flight that gives them the appealing air of fledgling birds, and perhaps this weak flight is a consequence of negligible predation in the period when the ancestral species was spreading out across the barren islands, and at a later time a cause of that isolation on the separate islands that permitted distinct species to evolve.

The tameness of Galápagos creatures is striking in these finches, which are thought to be the first land birds to have colonized the archipelago; the thirteen species here, together with the Cocos finch, which is peculiar to Cocos Island, six hundred miles to the northeast, derived originally from a single species that may have arrived in storm, though which if any of these species is the ancestral form, and which bird of the American mainland its closest relative, remains a mystery. The effect of innocence is much enhanced by their habit of coming confidingly to the most casual call, and seen so close, without the filter of binoculars, the little birds are luminous with life. Even if one has read about this innocence, and been prepared for it, the reality is very moving.

Followed like Saint Francis by my flock, I smile and talk to them, feeling something akin to shame.

The birds' tameness has survived several centuries of destruction of the island's fauna. Pirates and buccaneers, sealers and whalers, are thought to have taken hundreds of thousands of the great tortoises, which are now very scarce except in restricted localities on Isabela and Santa Cruz, and absent entirely from several islands where they formerly occurred; those races—four of the fifteen known—are now extinct. Meanwhile, domestic animals—horses, cattle, donkeys, pigs, goats, cats, and dogs—together with man's cosmopolitan companion, *Rattus rattus,* were turned loose on the defenseless native fauna and sparse flora, and did much to increase man's impact on the islands. As late as the 1930s, the tortoises were slaughtered for their oil on Santa Cruz, and the American servicemen assigned to Baltra diverted themselves by destroying most of its inhabitants, including the endemic hawks, mockingbirds, and dove, and also a race of the land iguana, which was common on Baltra when William Beebe visited here in 1923. A few species of finches, the yellow warbler, and the small lava lizard are the only vertebrates that have survived.

At 6 A.M. on October 28, the schooner sails northward toward San Salvador, known also as James Island and Santiago. To the sixteenth-century Spanish these islands were known as the Islas Encantadas, "shadows and noe reall Islands," because of their tendency to shift on the horizons, or so it seemed to early navigators beset by strong and uncertain currents, fitful winds, and melancholy calms. Subsequently they were called the Galápagos, after the tortoises that became the chief reason for visiting them. The English buccaneers who hid among the islands between raids on Spanish shipping gave them Anglo-Saxon names, but in this century Ecuador has decreed them the Archipiélago de Colón—hence such names as Fernandina and Isabela, Pinta and Santa Maria. Some of the islands such as Santa Cruz, the former Indefatigable Island, have as many as five names in common use.

Sea and air are silver under the dawn clouds. The world looks monotone and lifeless, but life is everywhere. Pelagic shearwaters and storm petrels mix with frigatebirds, masked boobies; the tiny storm petrels—there are three species here—cast faint shadows on the sea. The frigate-birds pick at fish schools driven to the surface by dorado, two of which, hooked and spanked aboard, turn from emerald-blue to gold, then gray. So much for supper.

Off to port, the dark triangle of a shark; silver dolphins come and go. The twin wing tips of a manta part the surface and, for a strange instant, as if tuned to the ship's passage, hold dead still.

The schooner anchors in the lee of Bartholomew Island, which is separated by a channel from San Salvador. This southeast region of San Salvador is a vast black expanse of pahoehoe or plate lava, probably laid down in this century, for it is still bare and raw. Here and there on the pahoehoe are scoria cones of varicolored matter spat up from the Earth's crust beneath the sea. Some are a demonic red, and white streaks of guano, like dead ash, give this shore the infernal aspect that so enchanted the romantic writers of the nineteenth century and not a few in our own day: the peculiar beauty of the Encantadas and the variety of its aspects are rarely mentioned.

New life that would break down the lava's tortured swirls to form a soil is discouraged by aridity. Here and there in nooks and fissures a sedge has taken hold, and a few small barrel cacti, *Brachycereus,* and there is a short-horned grasshopper with fire-striped legs that wanders out into the fierce heat of the lava and no doubt dies here.

Underfoot, a hollow metallic ring where lava bits fall off into the fissures, and a strange shine in the hard briny light; no sound in the world but the hard wind of the trades that drives wind and water at the southeast faces of these old volcanoes and will one day return them, with all of their odd plants and animals, to the sea. The sound of wind is a thin whistle in burnt stone; the wind itself is silent.

I climb a small volcano. The crimson throat of a female lava lizard, the solitary yellow blossom of a small composite, are seeds of color in a place where colors have been burned away.

On the rim rock, a hawk's shadow; it passes calmly, close over my head. This endemic buteo, a relative of Swainson's hawk of the Americas, is a miniature golden eagle in appearance, dark rich brown with bright yellow points of color in the nares and legs. It is the only native hawk except for the occasional osprey. The hawk circles the small crater twice before alighting on its nest halfway down the wall in a white palo santo tree that in this season has no leaves: this incense tree (*Bursera*) is a relative of the gumbo-limbo of Florida and the Caribbean.

A ground finch and a mockingbird, the hawk . . . three birds. This is enough. The lava lizard, changing its aspect and dim colors very little from island to island, is the most prominent life left in most of the Galápagos, once one has left the edges of the sea; other small reptile species are confined to a shy gecko of the highlands and a genus of small constrictors, *Dromicus.* Here and there the finches flurry; otherwise, birds, butterflies, even spiders and insects, are few. Due to the sparsity of plants and animals, one notices each individual, *perceives* it.

Old silver driftwood from the jungle coasts—Cocos Island, perhaps, or Panama. Such wood might come on the warm El Niño Current that moves southward in December as the *garua* comes to an end, bringing warm water and humidity, flying fish and morning rain. Much of the flora and fauna of the islands has affinities with Central America and the Caribbean, and may have come originally on El Niño. One such creature is the red crab, *Grapsus grapsus,* called here by its Jamaican name of Sally Lightfoot.

Sea lions loop along the shore, and a group of penguins shuffle in and out of the dark sparkle under a pinnacle that juts one hundred feet or more out of the sea. The penguin's pelage resembles fur, and it has the same shine as the sea lion, lustrous with fish oil and sea minerals. One waddles up out of the sea, gazes at man, and waddles down again, emitting a soft moaning. This most northerly of the world's penguins may have come here later than the other endemic birds, for it displays a sensible discomfort in man's presence.

Eared seals or "sea lions" abound in the Galápagos. The common species is a race of the California sea lion or circus seal and presumably arrived out of the north, while the uncommon one, much smaller and more localized, is a fur seal, kin to the southern fur seal of cold austral waters. Presumably the fur seal came north on the Humboldt Current with the penguin, which is very similar to the Peruvian penguin and the Magellanic penguin of the Strait of Magellan and Tierra del Fuego.

In early afternoon we make a dive on the weather side of Bartholomew Island. These pale seas and pale islands, cold water and sea lions, remind me of South Australia, a haunt of the great white shark, and I suggested at break-fast that this beast might occur here. Later in the morning I saw a shark jaw curing on the deckhouse; it came from the coast of Ecuador and had the triangular serrate teeth that distinguish the white shark. Captain Richard Foster, an enthusiastic diver who once worked on the abalone beds of California, says that local fishermen are familiar with a very large pale shark that comes boldly in to boatside after hooked fish, and as this is a known trait of the white shark, these clues make our diving an adventure.

The wind has risen, and the sea is roiled.

Around a lava head drift stylish parrotfish, wrasses, butterflies, tang, grunts, grouper, angelfish, and other families of tropical reefs around the world, though nearly a third of the species here are restricted to the Galápagos. Fifty feet down, the light turns to purple shadow, and thirty feet deeper lies a waste-land. All by itself on the somber ocean floor lies an odd canister, like a fat amphora, and as Richard motions me away, I recognize it as an unexploded bomb. Later he told me that the Americans on Baltra used some of these outer rocks for bombing practice.

From the peak of Bartholomew, Marchena Island is visible to the north and so is Pinta; these islands lie on the far side of the equator. Beyond the horizon lie Wolf and Darwin Islands (formerly Wenman and Culpepper) the small northernmost outposts of the archipelago.

The sun sets behind the pinnacle that looms over the ship, turning the dead rock incandescent, the blue sky to silver. Against the silver a black frigate-bird on long still wings hangs poised, heraldic, in a cross. At darkness, strangely, the pinnacle turns white. The pale tower rises out of the black water, and behind its bent head, the constellations pass in diadems of stars—Scorpio, the Pleiades, Orion. In the clear skies of the Islas Encantadas, so far from foul kingdoms of man, the stars fall all the way to the horizon.

Night passage. The equator is crossed after midnight, and at sunrise the schooner is passing down the northern coast of Isabela; ten miles to sea rises Redonda, a formidable sea rock, hoary with guano. Great flocks of the northern phalarope and Audubon's shearwater that are feeding along the current lines offshore pick up in gusts and blow along like scud. Boobies, bos'n birds, and frigates. By itself is a Cape pigeon, strayed north out of Cold Antarctic latitudes. These seas of the western archipelago are rich in nutrients brought by the

Cromwell Current, which comes from the west, one hundred feet below the surface, and supports the great plenitude of sea life.

Deep ocean comes all the way in to the land. The northwest point is rounded close inshore, and above the ship looms the stillness of a broken crater, walls dark in early morning shadow, rim haloed in sunny mists. The running silver sea, rummaged by tide rips, chews at its base, and white-bellied pelagic birds skirt the black and shining shore where one day silent seas will preside again. A manta ray leaps at the sun, slaps down, and ocean creatures swirl just off the rocks—a shark, more mantas, an ocean sunfish, a swift school of dorado, flashing sea blues and emeralds in the morning water. The floating carcass of a sea lion, head and shoulders shredded by the sharks, returns the energy of its rich matter to the ocean.

The schooner anchors off Punta Espinosa, on Fernandina, an island composed of a single mighty volcano that rises 5,000 feet above the sea; in its crater is a lake that contains an island. From Punta Espinosa, on the eastern coast, one gazes in all directions at volcanoes, for Fernandina sits in the half circle formed by the volcanoes of Isabela—Volcán Wolf, off to the north, then Darwin, Alcedo, Sierra Negra, and Cerro Azul. Fernandina is the one large island in the Galápagos that has not been infested by domestic animals or rats, yet giant tortoises, despite reports of tortoise sign, have never been recorded here. It is a dark and inhospitable mountain, girded with fields of sharp *aa* lava, and so far its few visitors have been unable to resolve the mystery.

At dawn on Punta Espinosa, three whimbrel come and go, and tattlers are common on the outer rocks, and the Galápagos ground dove, a pretty chestnut bird with bright blue eye ring and speckled wings, wanders the sand and mangrove in small parties.

On all the rock points, marine iguanas, still sluggish from the cool of night, lie in swarms like demons, heads raised to the gray dawn. These heavy-spined black lizards, more dragonlike than any others left on Earth, must have dismayed the early seafarers, low on water, who first investigated the black lava shores of this island universe, and doubtless offered inspiration for the purgatorial descriptions of the place. ("It is to be doubted," Herman Melville wrote, "whether any spot on Earth can, in desolateness, furnish a parallel to this group. . . . No voice, no low, no howl is heard; the chief sound of life here is a hiss.") The transfixed animals seem to grow from the dead rock on which they lie, the shade of which so resembles their own that one can approach closely without seeing them. One or two drag themselves off across the bodies of the rest, which remain motionless, heads turned toward the intruder. The congregation of weird heads is startling—the short rounded jaws, adapted for chewing algae from the rocks, give the faces a subhuman cast that makes them terrible. The red crabs that walk over them and the finches that pick them rudely for their parasites are ignored. Yet this iguana can move quickly when it needs to, and in the sea it moves like a crocodilian, legs tight, with a sweeping tail. Toward noon I swim out along the point, hoping to see the iguanas feeding underwater—they cling to the rocks with their long claws, and may stay submerged for an hour or more—but those that are swimming resist my companionship, and finally I return into the shallows, where a sea lion lets the waves roll it like a shining log on the black sand.

Later I dive off Punta Espinosa with Richard, who has seen big sharks here, a dozen at a time; on all our dives we carry spears as a precaution. We swim off the shelf into deep water, then follow the incline down to one hundred feet. In this heavy wind, the sea is murky, and I am content that we see no sharks, which are poor company in roiled water. On shelves of rock are giant scallops with red mantles and blue, and a variety of algaes, corals, bryozoans; the fish legions are led by multicolored sheepshead of several species, and a grouper of pure gold.

A land iguana—the animal from which the marine species presumably evolved—comes forward from the mangrove shadows to investigate my foot; it wears a benign vegetarian expression, and pays no attention to the running ants upon its eyelid. The land iguana has the long lizard jaw that the marine species has lost, and is a sturdy patchwork of rich browns and chestnuts.

Tide channels extend far back into the lava, occurring where the crust has fallen in, but the clear water in the pools is all but lifeless. A few red crabs and roaming seal pups, a pair of great blue herons, small fishes that flash quick silver signals at the sun as they nip sparse slimes of dark red algae from the stone.

Over the lava, shrieking, flies a hawk. It fails to spot the elegant small snake, black with gold banding, that is the rarest of the three endemic *Dromicus* constrictors. Though found on the hot lava, in the sun, the little shining snake is cool. It is as tame as other native creatures of the Galápagos and makes no effort to escape.

Alongshore, flightless cormorants stand here and there against the sky, observing the approach of man with eyes of cosmic blue. This ragged cinder-colored bird, largest and the rarest of the world's cormorants, is found only in a few scattered places in the west part of the archipelago. One broods an egg among the black iguanas; another bows, erects its tail, and squirts a long arc of uric acid at the morning sky. Although its wings are little more than stubs, the flightless cormorant has retained the family habit of extending them to dry upon leaving the water, to be ready for flight lost long ago because, on a coast where predators were absent and fish plentiful, flight served no purpose.

Twilight. The black cone of Fernandina, between fire-colored sky and blue-gray sea. Black birds of sunset cross the fire and vanish in the silhouette of the volcano.

At daylight the schooner crosses Bolivar Strait to Punta Tortuga on Isabela, where the sun rises behind Volcán Darwin. The long beach at Tortuga is broad and high, and tracks of green turtle are abundant.

Behind the beach, red and black mangrove creak in the hard wind, but on the black mirror of the swamp, the yellow leaves float peacefully, and small birds come and go. One is the yellow warbler, very similar to its ancestor on the mainland, and the other is the mangrove finch, a small gray-olivaceous species

31

that, like the better-known woodpecker finch, is among the few birds in the world known to use tools: with a twig it chivvies insects out of holes in tree limbs. This bird tried one hole with its bill, then snapped a twig from a nearby branch with a quick twist of its head, probed the hole, procured the occupant, dropped the twig, and began all over, inspecting the next hole with bill alone (though sometimes the bird will retain the twig for the next operation).

The ship returns around the northwest point of Isabela. Where currents meet the wind out of Bolivar Strait, the seas are rough, and through the bright tumult pours a troupe of bottle-nosed dolphin that follows the ship eastward under the leeward shore toward Punta Albemarle, arching out of one wave's pouring face, piercing another in clean slash of white. Riding the bow wave, they roll white bellies to the sun, spin, flip, and slide away again without visible motion of the tail. A calf veers with its mother, holding magically to a fixed point no more than two feet from her side; such precise maneuvers suggest a system of communication analogous to that which permits a dense fast-flying band of shorebirds to whirl and veer as one.

A frigate pirates a young booby, snatching the falling silver fish with a rushing stoop that tatters its black wings, and lilts skyward once more without effort. It is this effortlessness of frigatebirds, of sea lions and dolphins, that is stirring: the mammals, at least, appear to revel in the doing as the ultimate expression of *seal-ness*, of dolphin existence. Speeding before the prow of a strange ship carries the dolphin nowhere but off course, if course it follows, yet it comes racing through the seas to play.

From a hard rope hammock I gaze at sun and sails, the rolling rigging on the blue Pacific sky, the seabirds coursing back and forth across the bow. To journey in this way on an ocean afternoon, coasting an immense, strange region—this is rest. Along most of the many miles of Galápagos shoreline, man's absence is entire—no sail, no smoke, no lump of human habitation. The lonely outposts must be sought, and might well be missed entirely by a vessel unaware that they are here. Once or twice in many days a battered fishing craft is sighted—a tuna boat or lobsterman, with names like the *Venus* (Manta) and the *Maria de Lourdes*.

Rounding Punta Albemarle at twilight, the ship meets steep heavy chop out of the southeast, and there is a toilsome passage to James Bay. Some of the passengers are seasick, and some of the crew as well. The seas sweep across the bow, and sheets of water on the foredeck sparkle with luminescence, and in the blackness all around strange luminescent swirls betray the presence of night creatures. At 3 A.M. the ship drops anchor in James Bay, on the west coast of San Salvador.

At daylight on the hills behind the shore three goats are visible. This resourceful animal is the worst enemy of native wildlife, cropping the thin browse to the root; it will eat through the two-foot trunk of the tree cactus in order to topple it and munch the spiny pads. (This species of *Opuntia* has taken an arborescent form as a defense against the giant tortoise; on the island of Marchena, where no tortoises ever occurred, the same species is recumbent.)

In former days, the tortoises came down commonly to the arid shores to feed on *Opuntia* and other plants and to lay their eggs, but prolonged slaughter together with competition for browse from feral animals and predation of eggs and young by dogs, cats, pigs, and rats, has driven the last of these reptiles into the hinterlands.

Behind the beach at the north end of James Bay, two brackish lagoons are set about with black mangrove and buttonwood. Here a flock of forty-odd flamingo lives in company with a somewhat larger flock of Galápagos pintails; they are close relatives of the flamingo and pintail duck of the West Indies. The flamingos are a brilliant powder orange, and a mating pair forms a glorious tangle of long legs and bright feathers against the livid greens of the pond margin.

At Buccaneer Cove, old litter of pirate picnics may still be found, and at the south end of James Bay, low gaunt-eyed buildings of an abandoned salt works are melancholy monuments to man's attempts to justify existence on San Salvador, which is presently without human inhabitants; the land iguanas so common here in Darwin's time are also gone.

I am relieved to slip beneath the sea, where Richard and I are instantly surrounded by butterflies and angelfish. A heavy-tailed male turtle, almost black (in parts of the Pacific, the green turtle, *Chelonia mydas,* is so much darker than the Caribbean form that it is known as the black turtle), comes to inspect us, and so does a white-tipped reef shark, small and neat, and in a crevasse grows the gasping head of a moray eel. Mollusks, uncommon elsewhere, are relatively plentiful: live murexes lie scattered on the bottom, and I find a pretty auger shell and a deer cowry. Oddly similar to the bomb at Bartholomew Island is an amphoralike storage container of the sort brought here by the early sailing ships, perhaps cast overboard by the same men whose roistering voices echo in Buccaneer Cove.

South of James Bay, in deep tidal fissures in the broad benches of rock, lives a population of the fur seal, which, like the penguin, makes its abode in the cool shadows of sea grottoes here and there among the islands. Until almost exterminated by nineteenth-century sealers, it was apparently quite common, and probably it is still more numerous than its retiring habits might suggest. Like its much larger relative from the north, it is an eared seal or "sea lion," but this species is darker, with rich pelage, and shorter in the face. By comparison, it is shy, but if one slips quietly into the water, the lovely animals —eyes wild in the thick lenses that bulge glassily underwater—will describe arabesques of silver bubbles all around.

At 5:30 A.M. a small party is landed on the east shore of Isabela. We set out at once on a six-to-seven-mile trek uphill toward the rim of Alcedo Crater, 3,700 feet high. A donkey track traces a ravine uphill through sparse groves of palo santo and euphorbia. I break off aromatic twigs of palo santo and sniff them as I go. Just as the sun bursts the low clouds over Jervis Island, a vermilion flycatcher, like a miniature red falcon, stoops out of the sky of a fair dawn, and I take a happy sniff in celebration.

A vast silence on the mountainside. One notices the droning of a bee, for bees, like other living things, are few. The common sound is the flitter of the short-horned grasshopper, colliding with intruders in fierce mindless flight.

On the first part of the ascent, I speak my poor Spanish with a young Ecuadorian crewman, Victor Hugo Castro, of Isla Santa Cruz. He would like to leave the islands, though he is happy that I find them "beautiful," a word that I use for want of a better. Except for a few striking prospects, such as Sullivan Bay and San Salvador seen from the peak of Bartholomew, the islands are not beautiful, yet their spareness and simplicity and silence are very moving, and suit me in a spare and silent time in my own life.

Old droppings of tortoise and wild donkey. Walking alone I come up quietly on a small herd of donkeys; the animals, snorting, thunder away, their hooves resounding in small booms on the hollow mountain.

A danaid butterfly, soft brown with black markings, seems to be dying, and I collect it for a British entomologist aboard the ship. (Three months later he would write me that this specimen was a first record of *Danaus gilippus* for the Galápagos.) Later I would wonder if the butterfly were merely tame, if I had not taken innocence for morbidity. This solitary danaid, three small yellow pierids, and a miniature blue—the smallest butterfly I have ever seen—are the only members of their kind that I see all day. This count is exact: one notices each insect other than ants and flies, and even these are less in evidence than they are elsewhere on Earth. One crab spider, one cricket . . .

Excepting a white morning glory, all flowers on the dry mountainside are small. The scarcity of pollinating insects has suppressed tendencies in island plants toward showy blossoms, and discouraged the establishment in the islands of whole families of common plants such as mints and acanths. A purple amaranth, small daisy flowers of a shrubby composite, a minute hibiscus blossom of pale orange and—just one—a vivid yellow pea.

I am pleased to be uphill from my companions. Crew and passengers alike are a singularly pleasant group of people, but on shipboard we are much together, and anyway, I found out long ago that I prefer walking by myself. It is a pity to talk on mountainsides—much better to listen to the silence, and exchange energy with the world around.

In the last one thousand feet, where the climb suddenly steepens, there is a swale of heavy bracken and high bunchgrass, hard to penetrate; I traverse the slope to a point farther north, and follow a tortoise path to the summit. A cloud that will overflow the crater is coming fast from the southeast, and I run the last hundred yards, in time to see it strike the rim and overflow down the inner walls, obscuring faint steams of fumaroles that rise from the crater floor. In the wet season, which is brief and undependable, pools form in the crater floor where the tortoises drink by day and convene again at night, submerging themselves in the warm mud and water. But many months may pass with no rain at all, and in the dry season the pools vanish. Sometimes donkeys gather in this steam from the infernos, and leave their skeletons in numbers.

The crater is enormous, somber, silent. To the north, the Volcán Darwin. To the west, the dark rim of Fernandina.

Tucking its talons beneath its breast, the Galápagos hawk makes terrific stooping dives down the crater wall. Like many land birds of the Galápagos, it descends to the seacoast in search of food, and the marine iguana is probably the main item of its diet, but it is of necessity a generalized feeder, taking carrion and even fish scraps snatched from the surface of the sea. In its agility and style, this handsome buteo seems part falcon: at Punta Tortuga I watched one far out over the water, kiting around among the frigatebirds. When on the nest, this hawk stoops readily on humans, and may strike them with its talons.

Summit silence, a dry wind, the blue glint of the Pacific far below. To a *Scalesia* comes a small dull olive-brown bird, innocent of markings, that hunts like a vireo through the branches; this is the warbler finch, which presumably evolved before the advent in these islands of the yellow warbler, which might otherwise have filled its niche. To the same tree comes a pair of woodpecker finches, the celebrated species that, lacking the long tongue of the true woodpecker, may use a cactus spine or twig to goad its insect quarry out of crannies.

Walking north along the rim, I come upon a tortoise. It has a view of Darwin Crater, but tortoise and volcano are mutually oblivious; to be tortoise or volcano is enough. After a time of utter stillness, as if dumbfounded by the sight of my brown legs, the tortoise gets on with its slow chewing. Three years ago, in summer, I saw this animal's close relative half a world away to westward, on Aldabra, in the Seychelles Islands of the Indian Ocean.

"At the East Pass of Aldabra, the giant tortoises occupy a high limestone ridge that forms a rampart of the northern coast, and their grazing of the native grasses has created a lovely woodland pasture shaded by casuarina and set about with a tropical shrub of acanthus and loosestrife: the pasture overlooks the shallow reef ledge and the sea. Overhead, seen through airy casuarina, the frigatebirds from the rookeries in the lagoon cross the bright sky. . . . From the dancing shade the tortoises watched me—I saw perhaps a dozen altogether—the light glinting on the ancient gray metallic heads. Although these animals have long since disappeared from all the continents, and no connecting population occurs anywhere between, they are so closely allied to those halfway around the world on the Galápagos that both populations were once assigned to the same genus (Testudo).

"At midday, the tortoises lay like boulders in the filtered shade, but later they moved out to graze, arching small heads on long necks to eat the grass, and bumping over the roots and limestone with an odd hollow ringing, like footfalls on the far side of a cathedral. These contemporaries of the dodo have fat round feet like those of elephants, and bronze scales with burnished rings, like the age rings in a cross-section of polished wood, and their fossil bones are almost as old as the island of Aldabra.

"White-flowered caper grew, and amaranth and nightshade and verbena, and from the windy trees came the sweet ringing of a rail. The flightless rails resemble small, neat pullets, and their rich chestnut feathering, washed with green, is iridescent in the wind-bared light of ocean afternoon. The birds are tame, entirely trusting, as vulnerable to the sticks of men as the great slow tortoise. Under the clear gaze of such creatures, in this bright whispering wood, there comes a painful memory of Eden."

The tortoise watches me with its small eye. This race of the domed tortoises (the saddleback races are found on low flat islands where vegetation is more sparse) is confined to Alcedo Crater, which can claim the last tortoise population in the Galápagos that is more or less intact. An estimated three to four thousand animals live on Alcedo, one of which, an august male, emerges from a bush ten feet away to have a look at me. Each volcano on Isabela has its own subspecies, for in former times, before this land mass was uplifted, these volcanoes were probably separate islands, and their populations remain separated by lava flows which the animals do not cross.

The large males mount the much smaller females wherever they can catch them—the females seek to avoid these crude encounters—and the mating is facilitated by a concave area on the after part of the male's belly plate that permits a closer fit. In copulation, the triumphant male emits strange primordial bellows, while the female maintains her customary silence. (A frustrated male may attempt to mate with boulders, though whether he bellows at these times is not recorded.) In the wet season these creatures of short leg and heavy shell may drag themselves down the crater's outer wall three thousand feet or more to feed and dig nests in the arid scrub behind the shore; the return upward presents no hardship, for they have time, and enormous strength, and mere hills do not impede them.

A quarter-mile walk along the rim produces three more tortoises. Though all cease their slow browsing to gaze at me, none are flustered but the one that I pat companionably upon the carapace. It hisses wearily and shuts its hinge in languorous alarm, opening promptly as I depart and stretching its neck out to peer after man in slow, dim titillation.

When the sun is high, the ancient animals retire to dust wallows and thickets, and at night they do the same, for nights are cool. A few hours of each day are spent in browsing; otherwise they rest.

Daphne is an isolated cone between Santa Cruz and San Salvador, a round volcano thrust up from the ocean floor. Its rim and outer walls afford nesting ledges for the masked booby and the red-billed tropicbird, while the inside walls, as well as the flat white crater floor, are inhabited by the blue-footed booby. All three species, homing with fish for the yawping young, are pirated by the frigatebirds that hang poised like black weapons on the hot thermals rising from the rock.

Land birds live in the prickly-pear and palo santo that have taken hold within the crater, and the Galápagos martin, a form of the purple martin, flickers endlessly around the crater. But the white stain and guano smell, the weird hissing whistle of male boobies, remind one that Daphne is an ocean rock, a haunt of seabirds. From the heights, a gigantic manta can be seen flying across the deep blue surface, leaving a white wake; it does whole sequences of great slow backward somersaults, white belly rising like a moon out of the ocean.

The masked booby, with yellow eyes in a sharp black mask and bone white plumage, is a clean, striking bird, though its webbed feet of olive green cannot compare with the baby-blue rubber feet of its near-relative. And both birds lack the splendor of the red-billed tropicbird, the bos'n bird, which I am seeing for the first time on the nest. Most nests are hidden: the bird's fantastic tail streamers betray its presence underneath a ledge. One female drives her young far up into the cranny, then turns on the intruder with a wild ratcheting shriek, black mantle raised to swell the white delicate form, pulse drumming hard under sea feathers of a silken white, red bill agape like open scissors. I back away.

The tropicbird that I know best is the yellow-billed species of the Caribbean, but I have seen this red-billed bird in the ocean off Aldabra, where frigatebirds and noddy terns are also resident. Until today, it was always aloft, flying on swift falcon wings into the ocean sun.

A mangrove lagoon on northern Santa Cruz is a breeding ground of the green turtle, and large animals are here in hundreds. Five heads at once may break the still surface to gasp the sea sigh of the turtle and subside again, and mating pairs are common. The male is not deterred by the presence of a small boat, but clings to the hapless female with the aid of flipper hooks that slip over the fore edge of her carapace. Like the female of the land tortoise, the female green turtle is an unwilling partner in this bumpy liaison, and struggles to flee both boat and dogged partner.

Everywhere in the lagoon, turtle flippers part the surface, the dark triangles of locomotion—sea lion flippers, shark fins, manta wings, the black outer primaries of seabirds—that in ocean habitats are shared by all vertebrate orders. Mantas and sharks are common in the lagoon, but in the mangroves, life is scarce—two lava herons, an egret, a yellow warbler, a warbler finch. A brown danaid is noticed and a dowdy dragonfly. The commonest insect here is the water skater, *Halobates,* the only marine insect in the world, which is found in tropical waters almost everywhere except at test sites of the atomic bomb.

Outside the lagoon are steep white beaches where the turtles dig their nests. Here as at Bartholomew Island and Punta Tortuga on Isabela, the green turtle seems less cautious in the camouflage of the nest site than its kin on other seacoasts of the world; perhaps the endemic turtle shares the tameness of most creatures of the Galápagos.

A dawn sail before the wind to Jervis Island, due northwest. Sea and silence. Not having to transcend the engine, human voices lose their stridence and blend with the light wind and the creak of canvas and the white cascade of the bow wave on the black surface.

In a sea lion colony at Jervis Island, the harem bull or "beach master," defending territory, is aggressive. Entering the water with a scuba tank, I hold my spear point at his nose tip as he circles me too close; not long ago, a girl tourist had a big seal bite taken from her calf. I am nervous, for the shallows are roiled, and these animals are very big and quick beneath the sea, but once I reach deep water he ignores me.

I swim down the volcano side. At sixty feet, I drift in a mournful shadow world of the old lava that perhaps a million years ago boiled up through the Earth's crust. A big male green turtle comes and goes, and three sea lions, swirling like huge otters. Silhouetted in the fish schools in the silver sun far overhead, the lovely animals describe sinuous *s*'s and pirouette upright and shoot away again.

At Conway Bay, on the northwest point of Santa Cruz, a series of white beaches are set off by black rock points, and offshore at either end of the bay's crescent are silent islands. Eden, at the western end, where the schooner lies at anchor, is a half crater with a green olivine beach; the sea rocks called the Guy Fawkes Islands lie off to the north, with Jervis, San Salvador, and Isabela in shadows of shifting depth on the horizon.

Ashore, the sandy tidal flats among the mangroves are all but devoid of life. Behind the mangrove, in a thin mixed wood of prickly-pear and palo santo, with a few acacia, lives a population of the land iguana, no longer common in the islands. A dozen or more of these heavy animals, yellow-gold with a deep chestnut back and tail, lie outside their burrows in all stages of inertia, gazing with equanimity at their visitors. But two golden-spined males are fighting, and have drawn blood, despite the ritual nature of their combat, which mostly consists of leaning heavily against each other.

In the dry sun, the small gray woodpecker finch sits on a limb and preens without a care as man holds his breath ten feet away. Then it flies to another limb that lies against a cactus pod and struggles to withdraw a spine. The spine glints in the sun as it is twisted, but is not pulled free. The finch commences feeding on its own, scouring a palo santo like a nuthatch and pecking at hollow spots like a woodpecker, then knocking off bark to get at the larvae of a long-horned boring beetle. Nearby, ground finches, a cactus finch, and a small-billed tree finch hop up in turn onto a rock, from which they fly to a goat-gnawed wound in an *Opuntia* that is bleeding precious water.

Night passage.

Several sea lion colonies of forty to fifty animals each are crowded on Champion Island, a rock off the north coast of Floreana, and since their domains are less than a hundred yards apart, the beach masters bellow dismally at one another all day long. There is a nesting population of the beautiful swallow-tailed gull, which has a large dark nocturnal eye ringed with a crimson circle. Such shallows as are found on the Galápagos are all but lifeless, and this odd gull has adapted to the lack of onshore feeding grounds by foraging at sea, at night, when the plankton communities rise to the surface. The Champion Island mockingbird does not cross the few hundred yards to the large island of Floreana, which has no mockingbirds at all; feral cats are thought to have exterminated the Floreana population, but as such cats are common on other islands, and mockingbirds, too, the situation is somewhat mysterious.

Undersea, the island walls abound with sessile life—why here, not other places we have dived? Bryozoans sway in plantlike colonies in the current, and there is a whole collage of crusting corals, and outcrops of the valuable black coral at one hundred twenty feet, where I begin to feel a little stoned. But life persists below that depth in this islet of anomalies: a sheepshead, white with dorsal streaks of black and orange, is ghostly in the crusting coral where the slope of the volcano overflows into black mist.

We rise again, in attendant clouds of angelfish and tang. A green turtle, like a benign gargoyle, peers down from a garden ledge of purple algaes and yellow bryozoans: in these dragon islands, even below the surface of the sea, the reptiles watch a man come, and watch him go.

At noon we go ashore at Black Beach on Floreana, where a German family has a rough farm in the hills, and ekes out its existence with a pension and supply store; here small sailing craft, some of them bound west across the Pacific, may take on fresh eggs, fruit, and vegetables. Black Beach and Post Office Bay, where one may place homebound mail in an old barrel used by sailing ships since the eighteenth century—it was first set up in 1789 by the British whaler *Aurelia*—come as close as anything in the Galápagos to a tourist attraction, but they are scars on these long stretches of still coast, and the less said about them, the better. Since the advent of man has destroyed virtually all Floreana's endemic wildlife—it is now the poorest of the large islands in this respect—the chief interest of the place lies in the number of human beings who, over the years, have vanished here without a trace.

Loberia, near Post Office Bay, is one of a cluster of islets in a clear green tidal current. It is a rock garden of tree cactus inset in russet beds of the succulent *Sesuvium,* on a base of squarish lava cobbles unlike any formation noticed elsewhere. Soft setting sun in the gold spines of the cactus, the fresh green of red mangrove, glowing seal forms, scarlet crabs, blue ocean light.

A dry equatorial autumn. The clouds move steadily north and west, piling against the southeast slopes of the higher islands, curling free again to drift to the west horizon. On the brown islands, the cones laid bare by parching light stand forth from those still remote, mysterious, under their cloud shadow.

Sunset. A solitary frigatebird in a yellow sky over the northeast point of Floreana. Behind the black point, the tropic sun is melting the horizon. In the Marquesas, three thousand miles away to westward, it is now mid-ocean afternoon.

Toward dawn, the Southern Cross is visible, and with it, a truer and more handsome cross, called the False Cross by navigators because it does not point at the South Magnetic Pole.

At Hood Island in the far southeast, the female lava lizard has a red head, the mockingbird has a long beak and yellow eye, the large and medium ground finches, abundant elsewhere, are entirely absent (their niche is occupied by a *Geospiza conirostris,* the large cactus ground finch), the marine iguanas are cinder red along the flanks, and the waved albatross, which may wander as far west as the Sea of Japan, makes its only known abode. The Hood tortoise, now extremely rare, has developed long legs, long neck, and an open saddle-shaped shell that permits it to feed much higher on its bush than can the domed tortoises of Isabela and other islands (other dry and barren regions of the archipelago have also produced saddle-backed tortoises). The known tortoises on

Hood—fifteen in all—have been removed to the Darwin Research Station on Santa Cruz as the nucleus of a propagation program to be put into effect as soon as the goats on their home island have been brought under control, and there is hope for this population that a few years ago was doomed.

At Punta Suárez, at the west end of the island, magnificent black sea cliffs on the windward side round to a cove cut off from the sea surge by a point of rocks. Masked boobies occupy the cliff rim, with swallow-tailed gulls on the ledges below and the blue-footed booby behind. Both booby species, which have just begun their courtship cycle, are weeks behind those nesting at Daphne, seventy miles away. The bluefoots step up and down in place to display the indiscreet blue-blue of their plastic feet.

In the Galápagos, there is a six-foot fall of tide, and marine iguanas gnaw brown algae on the flat tidal platforms under the sea cliffs, clinging with long claws to a yellow crust of giant barnacles and flattening in the manner of the Sally Lightfoot crabs as the seas crash across the rocks. On a perch above I sit for a long time and watch them. One by one, as the tide rises, the creatures release their clutch on the outer rocks and come washing in under the cliffs, the black shapes small as nerves in the boiling white, but soon an impassive head, with its lips that look sewn together and its salt glands that spit white dragon brine from the snout, appears over a rim of rock as the lizard climbs. Later in the day, when the tide is high, I find one one hundred fifty feet above the sea, on the edge of a sheer cliff. This one is a courting male, gone sulfurous bright green along the dorsal and on the legs, like a bronze heraldic dragon.

The simplicity of the littoral community is very satisfying. Brown algaes and a red one, the giant barnacle, the red crab—no other invertebrates are discernible through binoculars. The marine iguana represents the reptiles, the swallow-tailed gull the birds, the sea lion the mammals, but other species come and go. A green turtle is floating just outside the milk-green outwash of the surf, flying downward every little while to graze the deeps, and lava lizards venture to the spray zone to ambush flies and beetles drawn to a dead sea lion, dead boobies. Tattlers, oyster-catchers, a lava heron, land birds—yellow warblers, mockingbirds, doves, and finches—all visit the tide lines in the common instinct that their sustenance must come ultimately from the sea. In symbiosis—they do this also with the tortoises—finches pick parasites from the marine iguanas. (On Wolf Island—and on Wolf only—the sharpbilled ground finches will pick at the elbow of a nesting booby's wing until it bleeds, then drink the blood.)

A warbler finch, a tiny tuft of gray among gray rocks . . . doves and mockingbirds poking in and out among the boulders. Here on Hood Island, or so it seems (I'm still unsure of my own eye here) the bill of the dove resembles that of the mockingbird, an atypical pigeon bill, not only in length but in sharp point and downward curve. Both birds are euryphagous foragers here, and are often seen feeding together; presumably they compete for certain foods. Both appear tamer than their relatives on other islands, and the Hood mockingbird is bolder and more inquisitive than any other: I wonder if the boldness and long nosy bill evolved together.

Afternoon on the windward cliffs, a hard breeze from the southeast. The ocean surge shines the black walls with brine and forces explosions of cold steam out of a blowhole. Salt mist rises to the rim and blows away over the island.

White guano rocks and a green-white roll where broken seas moil with the blue-black deeps. Sweet stink of sea lions. In slow motion, the red crab tiptoes from the bill of the still heron. Shearwaters hurry in upon the wind, as if to hurl themselves against the basalt walls; within inches of the rock, they veer, riding the wind curve in the great amphitheater of the cliffs and fleeing away over the sprawling wash. At eye level, the albatrosses pass, and tropicbirds are here in numbers, streaming seaward in parties of eight or nine, nineteen or more; over the ocean that stretches away southwest toward Easter Island, the white archangelic birds, one hundred at a time, are visible as far as the eye can see.

At the highest cliff-point is a rock turret where four hawks have an eyrie, and the scream of these birds is the solitary sound that pierces the sea boom and the rumble of black boulders. One stoops so close that its passage resounds as a buffet in my ear, for the other side of this bird's tameness in man's presence is its willingness to attack him. It turns back down along the cliff, then comes a second time out of the western sun as I wave it off with a brandished stick. The hawk settles nearby, regarding me askance over its shoulder.

A short distance inland, in a hard xerophilous scrub, the waved albatross lays its heavy egg among black rocks. Here and there stands a solitary *Cordia* with rare blossoms of bright yellow, the only wild plant with flowers of fair size (an inch in diameter) that I saw in the Galápagos. Many birds are tending giant chicks, now nearly fledged, but others are still courting. The male raises its tubed-nosed yellow beak straight to the sky, emits a moan, then engages in furious bill strife with the female; the bills are clacked together at high speed in a sort of knife-sharpening motion, all sound and no fury. Most of the birds merely walk out of my way, but one takes off laboriously, tripping over the low rocks. At the first waft of ocean air, it sets its eight-foot wingspread and glides away seaward like a missile, as if it would never flap again until the day, perhaps two years hence, when it would reappear out of the ocean wastes to westward and curve on its white wings into Hood Island.

Night passage. The Magellanic Clouds, over Antarctica. These soft puffs of white mist are satellite galaxies of the Milky Way, the only galaxies besides our own that may be seen by the naked human eye.

I stand in the bow, leaning back against the forestay, as Hood Island falls away on the starboard quarter. Daylight will find us in the harbor of Santa Cruz, where two days will be spent in visits to the Darwin Research Station and the cloudy country of the highlands. The research station, which advises the Ecuadorian government, has turned the tide in favor of the remnant animals of the Galápagos, and the visit will be interesting, yet I do not look forward to it, being happy in the spell of these still shores of the Encantadas that pass out of my life off to the southward.

We sail at daylight from Academy Bay, on a course east-northeast along the coast.

Seen from offshore, the green haze of the Santa Cruz highlands does not convey the density of upland vegetation on the windward mountainsides where weather gathers. But this luxuriance mostly springs from such exotics as bamboo, avocado, citrus, coffee, and good pasture grasses. The endemic woods of the archipelago are dry-leaved, airy, thrifty: there is none of that fleshy-leaved waste of life that oppresses the spirit on the equatorial tropics of the Amazon, less than a thousand miles due east. And as in the arid coastal lowlands, one remembers every bird—two Galápagos ducks, nine cattle egrets, a pectoral sandpiper, the woodpecker finch, a cuckoo, and the small black rails of the high grasslands, which, like the rails found on Aldabra and other oceanic islands, may eventually become flightless.

A change in the weather. The highlands are obscured in rain, though Barrington Island, ten miles away, is bright in the morning sun. Barrington's tortoise is long extinct, and goats were destroying what was left of the island flora when a control program eliminated them entirely. Some land iguanas and the endemic rice rat (not a true rat; it may be closer to the hamster) still survive there; everywhere else, except Fernandina, where no exotic animal has taken hold, the rice rat has been exterminated by *Rattus rattus*.

Another victim of the black rat is the dark-rumped or Hawaiian petrel, which is larger than Audubon's shearwater and makes a higher arc in its rise and fall over the waves. Because of the almost total destruction of its highland nests by rats and pigs, it appears to be the most endangered of all creatures in the Galápagos (it is also endangered in its only other known breeding ground in Hawaii). I thought I saw one between Isabela and Redonda, and here is another, veering across the trade wind like a boomerang toward the November coast of Santa Cruz.

Wind and black birds. Two tropicbirds, alighted, lift their shimmering white streamers in clean arcs over the water.

Phalaropes, dark-light, dark-light, over the whitecaps: the birds appear to increase speed where the bellies flash along the wind. West of Gordon Rocks, a host of shearwaters, with scattered noddies, flutter and pick at a long slick of fish oil. Nearer, a small, bright-red dead fish drifts past, not yet discovered by the frigates.

Under the barren coast of Santa Cruz, two islets commemorate a General Plaza. The schooner moors in the channel between them, and a party visits the south island, which rises from the channel shore to the seabird cliffs to windward. The east half of South Plaza is mostly barren rock, but the crevices and hollows overflow with *Sesuvium* in bright clumps of copper, orange, carmine, bronze, vermilion, with crusting lichens and stray portulaca and the strange delicate gray plant *Coldenia*. Here and there stands an arborescent cactus, bright bark afire in the morning light, and on the eastern point a yellow warbler in rufous cap hunts the damp boulders for crustaceans.

Goats have been banished from South Plaza, and at the western end, the vegetation has crept back from cliff ledges to form a sea scrub of *Maytenus*, *Zanthoxylum*, *Parkinsonia*, and *Bursera*, together with the tree form of *Opuntia*. This renaissance occurred in time to spare the land iguana, which inhabits the small island in good numbers. I interrupt a courting pair, and the neck-chewed female makes the most of the disturbance by hastening away. The golden male, for want of a better course of action, opens its pink mouth and closes it again as if struck dumb by disgust, then performs the curious head bob and shudder—perhaps analogous to the press-up of lesser lizards—that is this beast's most expressive means of venting its emotion.

Among the thickets, ground and cactus finches shelter their weak wings from the sea wind, and in the lee of a staff-tree, *Maytenus*, sits the pretty short-eared owl of the Galápagos, rich brown and streaky buff, with diurnal yellow eyes and imposing lashes. All about lie owl pellets and paired black wings connected by dry keel and wishbone—all that is left by this bird of careful habit of the shearwaters on which it gains its living.

The trades blow hard from the southeast, where Barrington Island drifts on the horizon. The sea is driven at the cliff, and in a storm light, in the shadow of those clouds that rain on the shrouded highlands of Santa Cruz, the sea is the deep oily green of the olivine stone in island lavas. We dive in the channel between islands, and again off the north side of North Plaza; the water is roiled by the hard wind. One octopus is dismantling another, and we see two small sharks and a third one of fair size; this is the Galápagos shark, pale sand in color.

Sea lions have humped uphill from the channel shore to lie in the spray of the windward cliffs, which gives their forms a sinuous silken shine. The sweet-dry smell of feces fills the wind, and centuries of excrement are encrusted in a glinting white patina on the smoothed rock paths that cross the island. One animal has moved its fishy bowel on the very cliff edge, and there a marine iguana, tail held oddly in an upward curve, as if in tension or distaste, is making a glum meal of it. Though nominally a vegetarian, the iguana is no less opportunistic than other island creatures, and will eat dead fish and carrion when these are handy.

Dead seal pups, white bone, scraps of red crab shell, skeletons of mollusk —cones, cowries, murex, and worn shells of the endemic Eye of Judas.

One yearling seal, fresh dead, lies not a yard from another pup that is feeding at the teat; it looks asleep but for extruded sockets in the head, for a gull has seized its eyes. The cow seals are solicitous and quarrelsome: neck-biting, one drives another into the sea in snarling dispute over their young, and gives her antagonist's offspring a smart nip into the bargain. Along the channel range the beach masters, barking even as they lift their dripping whiskers from the water and continuing as they subside again, as if to threaten all below. This dismal barking, which comes out of the distance as a strange subhuman cry, is a sound of life in all wild parts of the Galápagos.

A booby fishes off the rocks, its plummet lost among the whitecaps, and a pelican contemplates shining stones on the eastern point. In the afternoon, three tropicbirds work back and forth along the windward coast, and on the channel shore are waders—sanderling, whimbrel, tattler, and ruddy turnstone, one of each. Audubon's shearwater and the swallow-tailed gull nest in burrows and ledges, respectively, along the cliff face. I lie flat upon the ground to dodge the wind and peer down at the incoming shearwaters, which, as at Hood Island, play dangerously with wind and cliff, banking away at the last moment over the wash of seas. Far overhead, the swallow-tails play, too, hanging motionless for minutes at a time where updraft is balanced by near-gale. The eerie chirrup of the shearwaters is just audible over the sea sound, but the gull's weird rattle at intruders, and a wailing cry that seems to despair of any future for its kind, pierces the thunder up and down the cliff, where its lone egg, gray and brown splotched on a ground of buff, lies in splendor in a nest of bronze *Sesuvium*.

DIARY OF A WHALING VOYAGE, BY ROY VONTOBEL

On a voyage not of destruction but of learning, men deeply concerned about the fate of the sea's greatest inhabitants sail to the Arctic as winter approaches to find, count, and, they hope, tag a few of the North Atlantic's vanishing whales.

I remember a scene from the film *Moby Dick* when the *Pequod*, hove to far out at sea, awaited the reappearance of the white whale. The great beast had sounded after being harpooned, and Captain Ahab (who will always look like Gregory Peck to me) and his crew irresolutely scanned the surface, not knowing where Moby Dick would come bursting from the depths to wreak his vengence upon those who dared pursue him. It was a charged moment. A curious silence hung in the air as the *Pequod* drifted on a dead-calm ocean. Even the gulls winging low over the water made no cry, as if they knew where the monarch of the sea was lurking and, unlike the humans aboard their puny ship, could sense what was to follow.

Such fanciful thoughts pass through my mind on a cloudy afternoon in early October as I stand at the bow of a ship in the Gulf of St. Lawrence waiting, like Ahab, for a whale to surface (blow, as the whalers say). The ship's engines are silenced and we drift almost imperceptibly in a tight circle. Great black-backed gulls glide quietly by on the gentle breeze. Minutes pass—four, five, ten. Wavelets lap against the hull. Dr. Edward Mitchell, a Canadian Fisheries Research Board scientist from Ste. Anne de Bellevue, Quebec, is speaking softly into his tape recorder, describing the behavior of this whale, and his voice is the only other sound. Looking north across the dull-black water to the distant Quebec shore, I think of villagers going about their routine while we watch the sea. This is Thanksgiving Day in Canada.

Fifty yards off the starboard bow an explosive whoosh shatters the suspense. Under a twelve-foot spout of wafting vapor, a massive shiny back, mottled blue-gray, ponderously arcs out of the water. In the crow's nest the watch shouts, and at the bow activity erupts like a stalled film suddenly started. Mitchell is talking rapidly into his tape recorder now, while others take aim with twelve-gauge shotguns to tag the whale with a visible streamer. The eleven-inch metal tube, which will encyst in the muscle of the smooth back, is harmless —and crucial to Canada's whale research program. Stumbling over the slippery deck, I vie for an opening between the two gunners to get a photograph of this encounter. The animal is a blue whale, its estimated length eighty feet. The tagging attempt is successful.

We are now eight days out of Halifax, Nova Scotia, aboard the sealer *Arctic Endeavour*. She is one hundred eighty-seven feet and wooden-hulled, nine hundred gross tons, an able ocean-going vessel. As we steamed beyond the harbor mouth and first felt the long low swell of the open ocean, someone reminded me that the ship is a veteran of many years at sea from the Arctic to the Antarctic. Besides Captain Ian McTavish, only a handful of the sixteen crewmen can claim the same. One of them is Newfoundlander Henry Mahle, with some thirty years' experience as a whaling skipper in the North Atlantic and Pacific. He directs our whale hunts from atop the *Arctic Endeavour*'s pilothouse, watching—as we all do—for the telltale wisps of whale blows and issuing his orders to the man at the ship's wheel through a tube.

After a whale is spotted and dives, it is Mahle who must put the ship near the spot where the animal will surface, estimating its movements and out-maneuvering its attempts to elude us, guessing its very thoughts. Mahle offered his services to the Fisheries Research Board after the Canadian government's ban on commercial whaling ended the East Coast industry—and his source of livelihood—in December 1972. (No commercial whaling by Canada has been conducted off the Pacific Coast since 1968.) The ban, imposed because of dwindling whale stocks in Canadian waters, shut down three whaling stations located at Dildo and Williamsport in Newfoundland and at Blandford, Nova Scotia.

At midafternoon the *Arctic Endeavour* steams at half-speed through Jacques Cartier Passage to the north of Anticosti Island. Mahle is at his customary post, conferring from time to time with the man in the crow's nest or at the wheel. Whales seem to be everywhere. A quilt of stratocumulus clouds melts into tufts against a narrow ribbon of sunlight along the sharp horizon. It is a gray day, but visibility is fair to good. The water is fanned in patches with ripples and is nearly flat despite a mild Gulf swell. This information is noted for our whale census, which is the main objective of the cruise. Each day, as the ship follows a predetermined course, we record all the whales seen in a strip up to eight miles wide. In the evening the length of the strip is determined and an esti-mate of the numbers of whales per one hundred square miles is obtained. It is a simple though expensive method of research, and only one source of informa-tion for whale biologists.

Five of us huddle on the gun deck at the bow, hunching up parka hoods against a breeze that is gusting up, waiting for another blow. Mitchell con-tinues his taped account of weather and sea conditions and reviews the whales seen since morning. Was that last blue one of the pair we saw earlier? Did any-one see that little minke whale pop up again? The questions and data accumu-late on the tape cassettes, which I will transcribe in the evening. Someone brings up cups of steaming coffee from the galley.

While waiting—so much of whale-stalking involves waiting—I muse about the small land birds that come on board like lost wanderers and flutter about the rigging, sometimes miles out of sight of land. Today a tree sparrow; yesterday two juncos. They are out of place in the fishy seascape, reminders of backyards and bird-feeders. The tree sparrow hops along the railing of the gun deck four feet from me. He is on his way to wintering areas in the United States from breeding grounds in Labrador or farther north. An hour later I watch him fly erratically away. Although it appears otherwise, I am certain he knows where he's going.

My attention is drawn back to whales as Mitchell nudges me and a shout comes from the masthead. We look hard in the direction Mahle is waving—a blow about three miles off the starboard beam. Sluggishly, the *Arctic Endeav-our*'s bow swings around and we are off in pursuit again. Thus the afternoon wanes, and it turns colder. At 1750 hours we no sooner go below for supper when more whales are spotted, this time another blue and a pair of fin whales. It is dusk now, and the light retreats rapidly. To the west a flaming sun hangs sandwiched between the low cloud cover and the horizon. "Wait till we get a little farther north," I am told, "the sunsets at this time of year are incredible."

The sharp-angled silhouette of a gannet heading southwestward in the direction of the Gaspé Peninsula passes swiftly across the orange disk.

Over supper we tally the day's deck count: eleven blue whales, four fin whales, and two minke whales. Around the table there are grins of satisfaction as our companions look up from bowls of corn chowder. Later, each of us sketches our impressions of a rendezvous of four blues that we witnessed this morning. Two pairs of whales coming from opposite directions met, churning the water as they rolled and dove about one another in what seemed, to human eyes, a tumultuous greeting. During the commotion one of the whales breached a third of its length obliquely (spy-hopping, in the jargon of whalers). At Mitchell's insistence we draw such sightings, for every observation can help refine the interpretation of whale behavior.

Although I have been busy with a camera, getting whales on film is a challenge. Only in the last few years, with better underwater equipment and at great expense, have revealing photographs been taken in the wild of some of the larger whales. The results already have corrected some misconceptions. An example: Rorquals such as blues and fins sink after being killed, and it has been the practice of pelagic whalers to fill the carcasses with air to keep them afloat while being stripped of their blubber in the water (as formerly) or while being towed by a modern catcher boat to a factory ship. These fast-swimming whales are now known to be much slimmer than bloated carcasses led earlier observers to assume.

The blue whale is the largest animal ever to have lived, reaching nearly one hundred feet in length and weighing one hundred thirty tons. Even before it reaches puberty a blue whale weighs far more than the most ponderous dinosaur ever did. And it is easy prey for a modern whaling vessel. Between 1910 and World War II, 215,000 blues died in Antarctic waters alone, and after the war the carnage continued. By 1965, when the International Whaling Commission agreed to accord the species complete protection, its extinction seemed inevitable. Few people cared. Few people even knew what was happening.

Although we see many fin whales in the Gulf of St. Lawrence, these sixty- to seventy-foot giants are not numerous in the North Atlantic and possibly never were. Yet the fin and the closely related sei whale were the major species taken by the Canadian whale fishery. Efforts are being made to phase out the hunting of fin whales where it continues, mainly in the Antarctic, but sei whale are still taken in large numbers there and in the North Pacific. The twenty-five-foot minke whale, looking much like a dwarf fin, once was spurned by whalers because it was too small to be hunted profitably while larger species abounded. Its days of peace are over, however.

Late in the afternoon a pod of twenty or so gleaming-black pilot whales, their bulbous foreheads and sickle-shaped dorsal fins unmistakable even at a distance, lazed with indifference near the *Arctic Endeavour*. Pilot whales once were the source of a lucrative land-based fishery in Newfoundland; as many as 10,000 a year were killed in the 1950s, nearly decimating the local population. In more recent years less than five hundred were caught annually in Trinity Bay; in 1969, less than fifty. A better example of overkill would be hard to find.

Our eastward passage during the night has taken us well past Anticosti. At sunrise on October 9 we are hugging the mist-shrouded north shore of Quebec, heading for the Strait of Belle Isle and the northern sea. Lying half-awake in my bunk, I hope in vain that Henry Mahle will go temporarily blind this morning or at least forget the location of the whale buzzer. We have rigged an alarm system—worse than any alarm clock I've ever heard—which can be triggered from the bridge when a blow is spotted. A wire runs to each of the cabins, and my buzzer is right over my head. At 0730 it jars me awake; fin whales off the port bow. No time for coffee.

Counting whales can be exciting or frustrating. Today we would see six fins and another minke. Gulf of St. Lawrence waters, like those of the chill Labrador Current of the North Atlantic, are rich with the tiny crustaceans called euphausiids and with copepods and other plankton upon which the large baleen whales feed. Large numbers of whales summer along the continental shelf from Cape Cod to Labrador and in the large inland sea that is the estuary of the St. Lawrence River. Fins, blues, humpbacks, and minkes can even be seen from shore, which has prompted the Canadian Fisheries Research Board to establish a whale observation station atop a high hill near Les Escoumins, Quebec.

Dolphins—small, toothed whales—come swiftly through Gulf waters, sometimes in great numbers, to investigate ships in the busy traffic lanes. The shiny backs of pilot whales passing in small groups cut the surface in sharp outline. Farther up the estuary, near the mouth of the Saguenay River, where an upwelling of deep cold Gulf water is caught in the tidal flows of a shallow channel, the big white Arctic porpoise known as the beluga gathers by the hundreds and more. Millions of seabirds also nest on the Gulf. Constant ship-followers, herring gulls and black-backed gulls cross and recross the *Arctic Endeavour*'s course, riding the updrafts above the pilothouse or at the ship's wake, sharp-eyed for bits of garbage.

Yet this nearshore abundance is misleading, for in the ocean beyond the slope of the continent the whale populations are spread thinly, and any life except for birds seems scant or nonexistent. Kittiwakes and fulmars follow ships far out at sea, and storm petrels skitter over the water or dart like nighthawks against a bleak sky, ignoring the passage of any vessel. Like the whales, the oceanic birds are larger organisms in the food chain and attest to the teeming diatom pastures in the northern ocean. But to a casual observer they seem all alone on vast expanses of water that is constantly in motion, where swell after swell gives a powerful impression of the physical forces—waves, currents, salinities, temperatures, pressures, light intensities—that determine the nature of the biological communities that can survive here. To a human the open ocean is a desert like no other on Earth, as inhospitable to a man as a mountaintop is to a whale.

Fortunately, even large areas of ocean that appear as blanks on a biologist's chart provide valuable data about where whales are found and help determine the patterns of their migrations. While Henry Mahle directs our hunts, it is Ed Mitchell who must decide whether we will learn more by going through an "empty" area or one where whales have been reported, either in research findings or a whaling company's catch statistics. Thus hours are spent late in the evening discussing whaling activities and catches of previous years, the findings of other scientific cruises, and what is known about whale populations in the regions we expect to survey.

Though we speculate about the effect the Canadian ban may eventually have on the numbers of whales in the Gulf of St. Lawrence and coastal waters, it is much too early for any noticeable changes to have occurred. Whales reproduce slowly; with a gestation period of up to twelve months, a female blue whale bears only one calf every two or probably three years. Thus the age composition and reproductive potential of any given whale stock must be considered. Furthermore, despite the volumes written over the years on the biology of whales, much more about their social habits must be learned to determine how the density of animals in any given area affects their reproductive rate. Extinction is not simply the result of killing the last animal, but of severe disturbances in breeding cycles and a chronic reduction of the number of young produced each year.

Much of this kind of information previously was sought by poking through stinking carcasses at whaling stations, slicing ovaries to count the scars of pregnancy, extracting "plugs" of the wax that accumulates in the ear canal to count growth rings as one would count the rings of a tree, and searching for other clues to the animal's age and reproductive history. Such information is more revealing than a mere count of animals. Ironically, while no one questions the benefit of the whaling ban to the whales, it has forced considerable changes on the course of whale research.

In addition to population structure studies and the strip census, two other sources of information help scientists determine where the whales are and how many of them there are. Wherever whaling continues, a comparison can be made between the effort expended to catch whales and the number of animals killed. An increase in the time spent hunting whales is an obvious measure of the declines of various species.

The random marking or tagging of whales is a fourth important source of data. Until recently this was also dependent on the catches of the whaling industry, for the approximate size of the whale population being hunted could be calculated by comparing the number of whales tagged, the size of the catch, and the number of tags found and returned. After the shutdown, a tag bearing long, brightly colored streamers that can be seen from a distance was developed. Since 1966 Ed Mitchell has spent hundreds of days at sea tagging whales in the North Atlantic, from Davis Strait, Greenland, and Iceland in the north to the Caribbean Sea, the Cape Verde Islands, and the Azores. Scores of whales —many fins, humpbacks, and sperm whales, and lesser numbers of other species —are today swimming around in the Atlantic Ocean bearing these tags. A considerable number of tags have been recovered and have contributed to the growing knowledge of whale migrations.

Such is the focus of our conversations during the long evenings aboard the *Arctic Endeavour*.

Our census in the Gulf of St. Lawrence is taking us along a deep trench where the likelihood of seeing whales is greatest. By the morning of October 10 we have veered away from the northern shore, making a wide sweep around a gently crooked elbow of land which I can see only as a soft blue-green haze of low hills. From this point the shore stretches away northeastward, one side of the broad funnel that becomes the Strait of Belle Isle at the top of Newfoundland. Boreal forest, endless reaches of it laced with upland ridges of glacial gravels and mosquito-ridden spruce bogs, covers this eastern arm of Quebec Province and extends northward over Labrador to the semibarrens and eventually the tundra. Hardy Leif Ericson, viewing Labrador from the Atlantic Coast, called it a "land good for nothing" and sailed away. Jacques Cartier called it "the land God gave Cain." My craving to trek inland and see it for myself is great, but we will be little closer to it on this voyage than we are now. "There are no whales there," Mitchell says.

By nightfall the *Arctic Endeavour* is plowing north at full speed. This evening the drawings and accounts for the field notes are being turned out in grand fashion in the expedition's workroom under the pilothouse. Recollections are vivid with the day's events. Mitchell, momentarily ceasing a furious scratching of his sketching pen, grins as an argument is waged over an aspect of one sighting and over the relative merits of certain observers.

Today the dolphins discovered us. At noon a choppy sea sparkled in shafts of sunlight, making it difficult to detect anything breaking the surface more than a few yards from the ship. Wind, steady from the northeast, made my eyes tear. It was one of those waiting periods, for we were stalking a fin whale. Half a mile away the flashing bodies of several dolphins burst from the water and paralleled our course. Minutes later they were all around us, leaping, scattering a million jewels in the sun. One dolphin cleared the water by over ten feet.

There were two different species, white-beaked dolphins and white-sided dolphins, both about nine feet long, perhaps fifty or sixty in all. They sped to the bow, the white-beaked dolphins usually one at a time and the white-sided dolphins in small groups, zipping through the water in effortless spurts, turning sharply, rolling on their sides as if to see whether we were watching them. As they broke the water, blowholes opened wide, gasped in air, and closed tight before the dolphins plunged nose-down and leveled off for another thrust just beneath the surface. They tried to ride the bow waves, but we did not seem to be going fast enough to suit them.

A number of white-beaked dolphins gathered about the fin whale, following its movements, and actually helped us pinpoint the spot where the whale would blow. Then they seemed to suddenly tire of their play and in an instant were gone. An hour later we spotted four more dolphins under the bow, swimming slowly side by side, one three-foot youngster tagging the flanks of a large adult.

The sea state worsened steadily as the afternoon passed into dusk. Waves are higher now, five to six feet, and an ominous wind rips off the crests in long streamers of spray, but the gulls trailing astern seem unperturbed by the weather change. In our workroom the portholes are fastened securely. There is still enough light to see an enormous tabular iceberg, wind-worn and wave-worn and probably many years old, far off our starboard beam. In summer, hulking bergs drift south in the Labrador Sea and often enter the Strait of Belle Isle, sometimes grounding and breaking up. "Heard gale warnings for the strait on the radio," Captain McTavish comments tersely as he comes in to issue the evening's round of beer.

In the wee hours of the morning a violent pitching and rolling of the ship awakens me with a start; one especially sudden lurch has turned over the heavy desk chair in my cabin. All over the *Arctic Endeavour* things are crashing and sliding around. Lying on my back, my arms and legs ache from spreading them against the sides of my bunk to keep from being tossed out. The fury of the storm outside seems to increase. As the ship rolls on its port side—where, below, my bunk is situated—I hear the waves pass over the deck above my head. From somewhere amidships, over the shriek of the wind and the banging of everything loose on board, a deep menacing boom resounds with every roll.

I must get up. As I negotiate the blessedly narrow passageway, falling upstairs, walking at ridiculous angles along one wall and then the other, there is shouting outside. Cursing loudly, the mate pushes past me. Our port lifeboat, broken free, is cracking its sides against its yoke stanchions and threatening to smash a cabin wall. Several seamen, in the black wild night, struggle to secure it with wraps of cable. Returning below, I look in on a shipmate. He has been measuring rolls of sixty degrees by the curtains on his portholes, he informs me with scientific detachment while chuckling in mildly hysterical amusement. He is sitting amid piles of books, clothing, and other debris in the middle of his cabin floor.

October 13. Off the Labrador coast. A huge swell is running, though the sea is otherwise glassy-smooth, and we are drifting in a thick, bone-chilling fog. The raging gale of the night before last seems long past; yesterday was spent cleaning up the ship while taking on water and fuel in the tiny fishing village of St. Anthony at the top of Newfoundland.

In this morning's gloom we are waiting to spot a pair of humpback whales. Henry Mahle sounded the buzzer today at 0830, but he might not have seen the whales at all except that we nearly ran over them. I have heard that humpbacks will come right up to investigate ships and swim around and under them. Blue and fin whales occasionally do so also, especially young animals. When I charged out the side door to the main deck, we were nearly beam-on to the swell and in front of me was a wall of green water in which one of the humpbacks seemed to hang suspended. I could have spit and reached him, but I was too startled to get a photograph and stood gaping as he slowly sank beneath the ship.

So we wait. Two stubby-necked puffins, legs dangling, appear out of the fog blanket, sail past the bridge, and veer off, wings buzzing madly in their tiring-looking way of flight. The fog suffocates us, and it is distressing not to be able to see more than a hundred feet from the ship. Even within our little pocket beneath the blanket, however, there are hundreds of shearwaters and dozens more of fulmars, chunkier but scarcely less graceful oceanic wanderers. Astern, the slick sea is pebbled as if by rain by a school of North Atlantic mackerel. Fulmars, mackerel, and the great whales alike come to feed on the larger plankton animals—crustaceans, jellyfish, and tiny squid. Capelin, silvery six-inch smelt running south by the millions from summer spawning grounds in the High

Arctic fjords, also consume enormous quantities of plankton. They in turn draw flocks of birds, shoals of cod, seals, and certain whales such as humpbacks to the Labrador waters and the Newfoundland banks.

"Blow!" Ed yells, looking in the direction of the vanishing puffins. The broad black back of one humpback, about fifty feet long, arches high out of the water and its flukes come up as it noses down for its deep "terminal" dive, the last in a series of blows. The other humpback, much smaller, soon follows. The length of time a whale stays submerged—and the depth it goes to—varies by species, and depends also upon whether the animal is feeding, resting, traveling, or being chased. Dolphins and porpoises ordinarily dive for about five minutes at a time. Deep-divers such as sperm whales and bottlenose whales have been known to stay down for as long as ninety and one hundred twenty minutes, respectively, though an hour is more normal when they are feeding. Rorquals—blues, fins, seis, humpbacks—usually remain underwater for ten minutes, although they could stay submerged much longer. In 1958 a fin whale became enmeshed in a dragnet cable off Cape Cod and, after struggling to free itself for thirty minutes, drowned. Had it not been fighting for its life, it probably could have survived for another ten minutes.

We have no opportunity to time the dives of the two humpbacks today, for after their last sounding we have not been able to find them again in the fog. It is aggravating to know that they may be only a short distance away. For twenty minutes we drift silently and listen for whooshes of blows we cannot see. But they are gone, their curiosity apparently satisfied.

The combination of this curiosity, a relatively slow swimming speed, and a habit of sticking close to continental margins has made the humpback whale easy to kill. In all the world's oceans these singers of the sea have been over-hunted; in the North Atlantic, Newfoundland whalers cut their numbers to a few thousand by 1951. As with many other species, the size and whereabouts of particular stocks of humpbacks and the movements of specific herds are not fully known. Humpbacks probably number no more than 1,300 in the north-western Atlantic and, like the blue whale, have received belated protection from the International Whaling Commission.

October in the Labrador Sea is ordinarily a month of bitter winds presaging the long Arctic winter. Seventy years ago, the whaling ships that crisscrossed Davis Strait and probed the fjords along the coasts of Baffin Island and Greenland already would have passed by on their southbound, end-of-the-season voyages. The *Arctic Endeavour,* however, is sailing north, and today, the fourteenth, is mild and sunny. Against a crisp cloudless sky, a raft of eiders lifts off the water and, flying low, heads toward the coast to hunt for mussels in the shallows. Astern, two thick-billed murres, bleating and croaking softly, slide down a furrow of churned wake. Small jellyfish, faintly pinkish, float here and there. Kittiwakes swarm, bolder than ever, and seem on the verge of alighting amidst the freshly washed T-shirts and dungarees strung aft to dry.

The visibility must be at least four or five miles; it is a perfect day for sighting whales. Finding them, however, requires a combination of careful calculation and sheer luck. Moreover, this is only one of the difficulties faced by marine biologists, for the mammals of the sea go about their lives in a world that is dark and boundless. How does one follow them on migrations that cover thousands of miles? How many biologists have fantasized about living just a single day in the life of a whale, somehow attached to its bulk like a remora to a shark and able to accompany it wherever it goes.

Instead, the whales' world exists for man only in one's imagination, given substance by whatever information biologists have been able to wrest from a variety of sources. Because he works under handicaps that do not beset those who study other kinds of animals, the whale biologist is not so quick to turn his back on earlier literature. He can delve profitably into a great body of whaling lore—fact and fiction, scientific and poetic—for library shelves overflow with journals of explorers, whalers, natural historians, and a host of other authors from Pliny to Melville. And it is the sperm whale, ancient and mysterious ocean wanderer, wrecker of ships, that has intrigued man more than any other.

Not until late afternoon is a sighting made. Half a mile away the massive angular forehead of a sperm whale breaks the surface; its curiously slanted blow, distinct in the crystalline air, is dissipated rapidly in the wind. Twenty seconds later we spot another blow, then another, about twelve in all. The noise of the ship's engine is apparently getting too close for comfort, however. Like a huge shadowy log the beast sinks on a deep dive, flukes up, and disappears, leaving only the Kittiwakes that wheel over a widening swirl of water to show that it ever existed. In the manner of all the notorious rogues of whaling history—Moby Dick, Timor Tim, Newfoundland Tom—this sperm whale effortlessly eludes his pursuers.

The blow of a sperm whale slants obliquely, rather than spouting straight up as in other species, because its blowhole is situated by a quirk of evolution atop the left side of its tremendous head. This is one peculiarity among many remarkable adaptations developed through the ages, for the sperm whale is a different sort of animal than the mysticetes or baleen whales—the blues, fins, and their kin. It is an odontocete, a toothed whale, whose early ancestors took to the sea shortly after the Age of Dinosaurs. It is also a giant among its relatives, up to sixty feet long. Another common toothed whale, the bottlenose, is half its size, and the killer whale is slightly smaller yet.

Whalers easily identify sperm whales not only by their slanted blows or the shape of their bodies but by their behavior as well. Their social habits seem to be more complicated than those of other whales; bulls maintain harems or families of females and young animals, and numbers of these groups make up the great scattered herds of sperm whales. Though bulls occasionally stray to the Arctic or Antarctic, the herds are found mainly throughout tropical and temperate oceans, keeping far out to sea where they prowl the black abyssal plains and submarine canyons for giant squid.

Yet much about the sperm whale and its marvelous adaptations remains a mystery. For example, an accurate description of its muscles and internal organs is still lacking, even though hundreds of thousands of sperm whales have been captured over the centuries. In the North Atlantic, where they once were common, the romantic era of sperm whale slaughter got under way when a Nantucket shore whaler was blown far from land during a storm in 1712. On the high seas, he found that the rare "cachalot" was not rare at all. By the end of the eighteenth century, however, so many had been killed that it was unprofitable to search for sperm whales in the North Atlantic and the ships moved to the South Atlantic and on to the Pacific. (In 1846 over seven hundred American whaling ships set sail.) During these decades, ten thousand sperm whales were killed each year for oil and ambergris; some twenty thousand sperm whales are now caught annually throughout the world. In the North Atlantic, whalers from Iceland and Spain still take small catches of sperm whales near home ports; one or two may be taken off Greenland; and a few more are caught in the Azores, where the deeps are close inshore.

Southwest Greenland appears, from the sea, to be an awesome, forbidding place. It is a land of fjords, from Cape Farewell at the southern tip, where a three-knot coastal current rounds and flows northward, to the north of Disko Island, where the Great Greenland Icecap comes down to the edge of the sea. On October 18 the *Arctic Endeavour*, having crossed the Labrador Sea, plunges through heavy swells past looming cliffs of gray rock, their lower portions mantled with snow and ice, and their peaks lost in dense clouds. Up the larger fjords, Vikings managed during the short summers to eke out a perilous farming existence for a few centuries. Today the Danes have built prosperous frontier outposts here, towns like Godthaab and Julianehaab, exploiting a lucrative salmon fishery. Eskimos, the native Greenlanders, have always depended on the sea, and their settlements are perched within easy access.

During the warmer months the outdoor markets in the Greenland towns peddle salmon, large fjord cod, and Arctic char taken from the nutrient-rich waters. Villagers pick over piles of eiders, dovekies, and guillemots. Seals— harp, hooded, and ringed—are taken on offshore ice floes, and farther north the polar Eskimos from remote settlements spotting the coasts of Melville Bay and Inglefield Bay stalk walrus. Some whaling is done, mainly for beluga and the more elusive narwhal, for the larger whales are now scarce. With special permits, Greenlanders in rugged little chug-chug boats take only a few humpbacks, fins, and minkes each summer, hauling them up fjords to lonely beaches out of smelling distance of the towns to be butchered. The hunter who occasionally fastens onto a sperm whale from his small craft quickly regrets his impulsiveness and returns to more easily landed prey.

It is bottlenose whales that abound, however. At midmorning Henry Mahle spots a series of almost indistinguishable blows. Scrambling up to the flying bridge I can make out the last blow, droplets glistening, backlighted by the low eastern sun. These are medium-sized whales about twenty-five feet long; slicing through the waves, they remind one of giant black dolphins. Though they generally move in large schools, we spot only two, and they come fast across the bow and vanish just as quickly. Rolling dangerously in huge swells, we decide not to chase them. Practice in eluding a ship would do them no harm, however, for their chances of being pursued by whalers in the near future are great. Together with minke, killer, and pilot whales, bottlenose whales are now the quarry of a growing small-whale fishery in the North Atlantic and elsewhere.

Scots first began to hunt bottlenose whales in the Arctic in 1877 and the Norwegians soon followed, but the development of more efficient whaling methods in the next few decades ushered in a whole new era of hunting large whales, especially blues and fins, in the Pacific and the Antarctic. Now, with the ranks of the remaining great whale species so depleted, hunters are again turning their attention to the smaller cetaceans. Minkes are taken by the thousands in the Antarctic, largely by Japanese whalers. In the North Atlantic, Norwegian whalers have been increasing their minke catches over recent years, and since 1969 these same ships have been scouring the ocean for bottlenose whales, from the Labrador Sea north to Davis Strait.

It is for these regions that the *Arctic Endeavour* now sets course, crossing the bottom of Davis Strait toward Baffin Island. By late afternoon on October 19 the ship nears the mouth of Cumberland Sound, altering course slightly now and then to stay well away from icebergs. To the north a sheer wall of rock is faintly visible, the craggy headland of Cape Mercy. (Names of many places in the Arctic, like Welcome Sound, Disappointment Point, Cape Hopes Advance, and Resolution Island, tell much about the hardships faced by early explorers.) The sea is smooth and the day, my first in the true Arctic, is overcast but mild.

Walking aft I find a petrel lying bedraggled near the hatch, the cause of its death mysterious, though I feel that somehow we are to blame. All around the ship are hundreds of dovekies—"bull birds," Henry Mahle says they are called in Newfoundland. They bob amidst bits of seaweed, dive for crustaceans, or scuttle along the surface, flapping short stubby wings and chattering in high-pitched squeaks. These little auks, which are occasionally pestered by marauding fulmars and kittiwakes, are one of the most abundant of northern seabirds and are caught by the thousands by Eskimos using long-handled nets. The dovekies seem unperturbed by our passing; nor does our intrusion offend a red-throated loon that is spearing small fish and wagging its head. To me, however, it seems that we have entered a world where our presence is disruptive and unwanted. It is the same feeling one gets upon blundering into a quiet library and seeing heads look up in silent but emphatic reproach.

Feeling strangely apologetic, I head up to our dayroom to type some field notes. It is dusk now. We pass scattered pans of floe ice, and several swimming harp seals stop every few yards to poke their heads up and observe us. Beyond them I spot something else and hastily grab binoculars, thinking that I have seen some rare, beautiful northern creature. It is an oil drum, bright red, floating by.

All the next day we steam steadily up Cumberland Sound, wherever possible skirting broken pack ice and, where not, trying to follow narrow leads. More seals watch us with benign interest. During the night, even with spotlights and our speed cut to a minimum, it is difficult to miss chunks of floating ice. One pan, 30 feet in diameter, bumps the port side; a line of polar bear tracks across the ice seems to lead right up on deck. Sheets of shimmering aurora borealis, with just the palest tint of rainbow colors, hang over the icy sea. Every now and then they move like gigantic curtains being jerked.

In the morning the *Arctic Endeavour* anchors in a small cove beneath the towering mountains of the Baffin mainland. We have reached Blacklead Island, a low spit of barren rock halfway up Cumberland Sound near its southern shore. Under a light covering of new snow we can make out a collapsed roof and the blocks of several building foundations, all that remains of a Scottish whaling station and a mission that administered to the local Eskimos employed by the whalers. From land stations such as Blacklead, Kekerten Island on the north side of Cumberland Sound, and Cape Haven farther south, the Scots hunted the bowhead whale until shortly after 1900. By that time, the bowhead had become rare in the waters of the eastern Canadian Arctic, Greenland, and northern Europe, for the combined depredations of Scottish, English, American, Danish, Dutch, and other whalers had reduced the species to virtual extinction. Bowheads can still be found in the Beaufort Sea, where they are hunted by Alaskan Eskimos, but hope for the recovery of the eastern population is dim.

Aboard the *Arctic Endeavour* a map has been prepared marking bowhead catches over the past hundred years and a course is plotted to pass through these places. Perhaps we will find a few bowheads migrating southward down the Labrador Current in advance of the winter ice pack. Perhaps.

The story of the bowhead whale is a sad one, but no more so than that of its close relative, the North Atlantic right whale. Right whales—and bowheads—are forty-five to fifty-five feet long, slow-swimming, and thus easily caught. Their oil-bearing blubber is by far the thickest of all whales' (that of the exclusively Arctic-dwelling bowhead averages twenty inches). Furthermore, their huge arched upper jaws carry the longest baleen or whalebone of any species. In the days when steel and elastic were unknown, this baleen was ideal for making whips, stays, umbrellas, and numerous other products.

Basques living along the Bay of Biscay began to hunt these animals in the eleventh century, and as the demand for baleen and lamp oil increased, their ships eventually spanned the Atlantic. (In 1578, a fleet of thirty Basque ships is known to have lain at anchor off Newfoundland.) By 1600 the fact that the Arctic regions north of Iceland and Scandinavia were teeming with right whales and another similar animal was common knowledge in the whaling industry, which was growing by leaps and bounds. In the 1700s European whalers began to shift their activities toward Davis Strait and Baffin Bay. Americans soon joined the hunt. Right whales and bowheads were pursued with better and better ships and equipment until, by 1900, they were no longer worth the trouble.

In 1912 only one ship left Dundee, Scotland, and it returned empty-handed. The land stations along the Baffin coast were abandoned. The bowheads were gone. Whalers bound for better hunting grounds far to the south only rarely happened upon right whales. There are no more than a few hundred of their kind left in the North Atlantic today, and the right whales of the southern hemisphere and the Pacific are scarcely more numerous. They are the survivors of one of the most incredible animal slaughters in history.

A misty dawn. Once again the somber cliffs of Cape Mercy, fringed by ice. Headed north, the *Arctic Endeavour* ploddingly rounds the Cumberland Peninsula, gently bucking a mild swell. Although the barometer is reacting favorably, anxious eyes scan the horizon for signs of a storm; a spell of bad weather in heavy ice could be troublesome. Other eyes hopefully search for a whale blow in the distance—the V-shaped spout from the double blowhole of a bowhead. To the west the shrouded land rises abruptly to heights of thousands of feet. To starboard a massive jagged iceberg looms, atop it the dark form of a large bird of prey, a gyrfalcon perhaps or even a sea eagle wandered far westward from northern Europe. To steal a plankton morsel, a glaucous gull harries the kittiwakes swarming over the upwelling water around the berg and over the turbulence of the tide races.

At midafternoon, near Exeter Sound, the engines are cut for an hour and everything is turned off except for one generator. The idea is to see if, by plowing ahead full tilt, we are scaring off wildlife. A troop of a dozen harp seals visits us, their bodies glinting with a sinuous luster as they roll in the dark water. It is an idyllic scene, but in the stillness the mutter of the generator sounds like a roar. Here and there are a few "Labrador poppies," small chunks of ice sculpted into elaborate forms by the slap and wash of waves. Pure-white ivory gulls appear, hovering for a few seconds amidst the seals, snatching at tiny crustaceans. We see no whales today, bowhead or otherwise. Late in the evening we cross the Arctic Circle.

Widening pack ice rims the coast, drifting slowly southward from the fjords and channels of the High Arctic islands. By late November or early December the floes will reach northern Labrador. Through leads in the ice we continue to probe the coast, searching for bowheads. On October 24, in a small inlet far up Ekalugad Fjord, we paddle ashore in the dory and dig under the snow to feel the coarse tundra grasses. Being at sea for a long time can make a man do strange things. A raven, picking over the remains of a caribou, is startled and indignant at seeing two men running toward him, yelling furiously. Simply to run on land is a joy. Earlier today I discovered an untapped store of apples in the forward hold, and ate half a dozen because they reminded me of warm summer days on the farm.

The bowhead search is discouraging. On October 25 we are on a course for Pond Inlet at the top of Baffin Island. The weather is worrisome, however, and no one is sure how much farther we can go. The *Arctic Endeavour* is running head-on into a good-sized swell, and on deck men are chopping away the ice that is beginning to coat the rigging. At midnight I poke my head into Ed Mitchell's cabin. He is poring over a map, and with his finger he is drawing a line north and west through the High Arctic islands.

Postscript: On October 26, rough seas, snow, and increasing pack ice imperiled the *Arctic Endeavour*. By supper Captain McTavish had made his decision to turn back. We had reached our most northerly point, some one hundred miles south of Pond Inlet.

In retrospect, our elaborately planned return course seems only to have served to bolster flagging hopes of sighting bowheads or other Arctic whales. On October 30, after zigzagging down the West Greenland coast, we put in at Godthaab to take on fuel and water. During a two-day stopover, three salvageable minke whale skulls were collected for the Fisheries Research Board from nearby beaches where they had been left by Eskimo hunters.

A week later, having crossed over to the mouth of Hudson Strait, where we hoped to find bottlenose whales, the ship was forced to hide in the lee of Resolution Island for thirty-six hours, waiting for gale-force winds to subside. Bad weather followed us down the Labrador coast. By November 9 we were back in the Gulf of St. Lawrence, where our voyage so auspiciously began. On November 12 we entered Halifax harbor, having seen no whales for three and a half weeks.

ELF OF THE ALPINE, BY ROBERT BELOUS

The exploiters of wild places—the clear-cutters, strip-miners, dam-builders, the purveyors of motorized, plasticized recreation—often claim that wilderness preservation is the goal of an elitist minority, that we cannot afford to "lock up" vast areas because a few people feel a need for solitude amidst towering trees and untainted, unfettered streams. But what, one might ask, of the needs of those creatures for whom solitude means survival?

It was cold but windless as twilight overtook my snowshoe trail near timberline in the Washington Cascades. I paused in a small clearing to gather some of the rewards of an alpine idyll—a craggy summit turning tawny gold, the muffled murmur of a brook from beneath its quilt of snow, the fleeting fragrance of spruce and fir.

It was only when I made a wide turn to leave that I caught a movement in the deepening shadows. A fan of spruce boughs lying across a log had parted. Thrust boldly through, framed by the bluish-green needles, was a furred and puckish face, its dark and piercing eyes fixed on me while its nose scribed a testing arc.

My audience of one had a head somewhat smaller than a domestic cat's, with eyelines that seemed almost cosmetic giving it the look of an elf. Its throat was splashed with pale orange, its dark feet amply clawed and almost too large for their owner.

A cautious step forward. I could now see a sleek coat of chestnut, blending with a buffy belly, flowing over an arched spine, ending in a burst of luxurious tail. Now only a few feet separated us as I remained motionless. And then my alpine elf took eight long and easy bounds, suggesting frightening speed, to a perch on a spindly sapling, uttered a soft and curious purr-growl, sprang to the base of a tall spruce, shot upward in a spiraling blur as though in pursuit of prey, shot down in an even faster descent, bounded like lightning to another spruce, and vanished into the topmost branches as claws crackled on bark.

Silence.

Then a lofty branch on another conifer bent under sudden weight. Claws raced. And a more distant branch swayed under a catapult, under the antics of the svelte aerialist of the North. For too short a moment, I had been witnessing that beautifully furred but seldom-seen carnivore which the French–Canadian *trappeur* called *le martre*. The marten, the Hudson Bay sable, *Martes americana*.

My elf was a cousin of the weasels, the mink, the fisher, the badger, the wolverine, the skunks, the black-footed ferret, the river and sea otters—the North American tree of the Mustilidae family, named for *Mustela*, the mink. And all these relatives share a most excitable temperament, insatiable curiosity, agile, muscular bodies. Indeed, pound for pound, they are the finest hunters among all the carnivores of our Northern forests.

In size, the marten is much smaller than the fisher and only slightly larger than the mink. An adult marten will weigh two to four pounds and reach a length of perhaps twenty-five inches. As with other mustelines, the male is larger than its mate. In comparison with the better-known mink, the marten has broader and more prominent ears, longer legs, and a more elegantly endowed tail. The marten's choice of habitat is the upper, alpine reaches of mountains and high conifer forests, though it descends to comparable country at low elevations in the Far North. The mink, instead, seeks out the watercourses of low valleys, ponds, marshes, swamps.

61

The mustelines are the "musk-carriers," and the marten has its pair of anal glands containing a viscous and pungent fluid. Though somewhat repellent in odor, the marten's musk hardly approaches the mephitic essence of the skunk. Nor can it be sprayed, though fright will usually prompt a scant discharge. A much more practical gland is located just under the skin of the marten's belly. Its fluid, rubbed on logs and roots along hunting trails, signals the marten's presence and marks its territory.

The energetic pace of this carnivore's life, and the attendant metabolism, means a constant search for food. If not asleep in its den, then the marten is surely hunting. Nor does it know hibernation during the long, bleak Northern winter. Then, the marten's feet become thickly covered with hair—offering insulation and furnishing snowshoes of a sort that let the elf bound easily along, even in the deep, powdery snowdrifts common to the eastern slopes of the Rockies. It leaves behind its characteristic two-by-two tracks, left and right prints almost touching, one having a slight lead, hind feet registering almost precisely in the front tracks, all faintly bearing the five-toe signature of the weasel tribe.

Winter, too, brings the marten its finest pelage. The shades of fur may range from blond to black, but the usual color is a lustrous mahogany brown with glistening guard hairs adding a counterpoint. The flosslike fur is soft and, like silk, cool to the touch. Darker fur on the tail and feet, plus that throat patch, make the marten one of the most beautiful furbearers of our continent.

That fact has meant its grief for many years. For the "American sable" was a favorite fur among the fashionable of eighteenth-century Europe, and the expert hunter became the hunted. In 1743 Hudson's Bay Company bought more than fourteen thousand marten pelts, and that was but the beginning. This curious creature was easy to trap and easier profit, and the only limit to the slaughter was the skill and greed of the trappers—and the marten's fast-declining numbers. By the 1920s, fur fads and scarcity had driven prices to $20 and $30 a skin.

Today, though, we know and appreciate the marten's role in the ecology of the North, and we can offer it the protection of sound wildlife management. Too, a small but growing industry is now breeding marten commercially. The one economic roadblock—a 230- to 260-day gestation period—has been cut in half by controlling the fur-farm animals' exposure to light.

The matter of marten mating in the wild was left to conjecture for years. Indeed, the marten is a solitary hunter with such savage intolerance for other animals that two marten invariably equal one fight. Yet there is a brief interval of grudging sufferance between the sexes, and on this, survival of the species depends. The mating season is from mid-July through August. The female mellows in ferocity, climbs in the alpine twilight to a not-too-high limb, and clucks or chirps to beckon a mate. When the two first meet there is suspicious maneuvering, taunting, a gauging of intentions, and several more-than-mock battles. If the female has not reached the proper stage of her oestrus cycle, she fends off the male with measured vigor. A too severe rebuff may send him into hasty though temporary retreat.

When the female is ready, subtle changes signal a new mood—a demure stillness, languorous preening, a growl that is evocatively close to a purr. The male's reaction is swift—but hardly gentle. He seizes the nape of her neck, sinking long and needlelike canines deep into the loose skin. Though she may occasionally give a low growl, he remains locked in this embrace for as long as an hour. Then both resume their solitary lives.

At this point, it is the female marten to which mammalogists look with interest. For despite her diminutive size, her gestation period exceeds that of a moose! Scientists now attribute this to a phenomenon known as delayed implantation. At the start the marten gestation seems quite average. But as the fertilized ovum approaches the stage of uterine implantation, all development halts abruptly. And for the next seven or eight months of harsh winter, this tiny speck of life lies dormant, awaiting a signal from one of nature's mysterious clocks. The sign is the lengthening daylight of February and March. Normal gestation resumes as though it had never been interrupted. A mere twenty-seven days later, the three to five young marten are born.

If they survive their first five blind weeks, the young marten finally view their world from a hollow snag or, most commonly, an expropriated squirrel's den high in a conifer—and comfortably lined with the fur of the previous tenant. The young marten quickly become hyperactive—swarming, squealing, cavorting in, out, around the den with such limitless energy that an onlooker might believe there are three times their actual number. By the age of three months they have acquired their lifelong taste for, and the skill to catch, plump rodents. And by late summer they will have scattered to solitary marten life in the alpine community.

The food of the marten varies with the season. Grouse, ptarmigan, and their eggs and nestlings are taken in spring. Berries, insects, pine nuts, and honey are eaten throughout the summer, as are the occasional rabbit, pika, and young marmot, as well as a few frogs. But the staples in the marten diet are the small rodents—mice, voles, woodrats, chipmunks, squirrels.

Nature artists often portray the marten poised high on a conifer branch, glaring with defiance over the limp body of a chickaree. But this picture may be more dramatic than prevalent. For careful field studies show that the single group of mammals most frequently preyed upon by the marten—one which outnumbers all other prey combined—is the Cricetidae family—the mice, rats, voles, and lemmings. And the single unrivaled favorite of the marten is the boreal redback vole, *Clethrionomys gapperi*.

But a catalog of biological data does not a marten make. Indeed, to those who know him, the marten is *the* delight of our alpine forests. Rangers in our national parks and national forests often tell of an elfish visitor, usually with a sweet tooth, that made "wintering over" at some remote cabin a happier stay. I know of one ranger wife who, throughout one long winter, looked forward to occasional tugs-of-war across a kitchen windowsill, the prize a dishcloth.

The fact is that where marten can live unmolested in undisturbed forests, their attitude toward man becomes an unusual mixture of curiosity and unalarmed wariness. A few individuals may even grow reasonably tame through frequent contact with people—but they will not suffer being handled.

Personally, whenever I have seen a marten in its sleek winter coat, darting through spruce at 30 degrees below zero while I stood swathed in layers of goosedown, gloved, snowshoed, hooded, hatted, snowsuited, booted, and bandoliered—and virtually inert—I have had to think twice about my ideas of "freedom."

If Americans succeed in preserving those few remaining primeval places where the marten can continue its timeless solitary hunt, this elusive elf of the alpine will always remain fiercely independent, courageous, forever free.

Audubon's Photographers

Portfolio One

Colorful dragon of the Galápagos Islands, the male land iguana is golden yellow with chestnut spines. (Photograph by Les Line)

*A giant tortoise drinks from a muddy pool
high on a Galápagos volcano.
(Photograph by Tui A. De Roy)*

Its jaws scarred from crushing turtle shells,
a big bull American alligator yawns in the sun.
(Photograph by Frederick Kent Truslow)

The tusks of the walrus are used to dredge up
the mollusks that form the bulk of its diet.
(Photograph by Leonard Lee Rue III)

The coat of a newborn harp seal is whiter than the ice of the Gulf of St. Lawrence. (Photograph by Franklin Russell)

Portrait of a coyote. According to legend, an Indian hero's son would be dressed in a robe of coyote skins to give him the animal's keen senses of smell, sight, and hearing. (Photograph by Glenn D. Chambers)

On the tundra of Alaska's Mount McKinley National Park,
a red fox carries a ground squirrel to its den.
(Photograph by Robert Belous, National Park Service)

TRAVELS AND TRAVAILS OF THE SONG-DOG, BY GEORGE LAYCOCK

Loved by conservationists and hated by sheepmen, the coyote has not only survived but prospered in the face of an all-out war of extermination waged by government agencies and their rancher constituents. Now, in response to a growing environmental ethic, the indiscriminate weapons of the predator poisoners are being set aside in favor of détente with this remarkable wild canine.

Some years ago trapper Bill Pullins arrived at a ranch in South Dakota to dispatch a coyote that had learned to eat sheep. Pullins, a rugged outdoorsman, takes pride in his knowledge of wild creatures and in his trapping skill. Surveying the landscape where the coyote lived, he found fresh tracks on a much-traveled trail between twin buttes. "I figured that would be the place to catch him," Pullins recalled. "I didn't think it would be any big job. But I was wrong." For week after week the two wise old-timers, coyote and trapper, waged a battle of wits. And Pullins needed all the tricks he had learned as a government trapper, or district field assistant as the job was later called, for the U.S. Department of the Interior.

That first morning, Pullins spread a canvas to kneel on, dug a shallow trench in the trail, and set a trap at nearly ground level. Next, he sifted loose dirt over the trap, putting aside the excess to dispose of somewhere else, and dusted litter and grass over the spot to make it match its surroundings. Finally, as a lure he sprinkled coyote urine on a nearby bush. Making the rounds of its territory, the coyote was expected to investigate the strange scent and place a paw on the trap's sensitive pan.

Pullins' trapping technique has worked well over the years. "I could have told you how many coyotes I've killed," he said, "but when it got to twenty-four hundred, they gave up the quota system, and I quit counting." During one eight-month period he killed two hundred twenty coyotes, mostly with traps, which he finds more challenging than shooting or poisoning.

But this time he had met an experienced coyote that knew about traps. "It seemed like a game with him," Pullins said in admiration. When he returned to inspect his first set, there was no coyote. But one had been there— the trap was in plain view and had been sprung. "He had dug underneath very carefully and lifted the jaw of the trap." Time and again Pullins reset the trap, only to return and discover that the coyote had dug it up and set it off.

Unable to get the coyote out of his mind, Pullins devised a new plan. He shot a prairie dog, placed it partly under a flat rock near the trail, and skillfully buried four traps around the rock. Later, he found every one of the traps had been dug up. "Besides that," he grinned, "the prairie dog was gone."

After six months of effort, Pullins finally did catch his nemesis. When found in the trap, "The coyote just looked up at me like he was saying, 'I sure pulled a dumb one that time.'"

Pullins, now retired and living in White River, South Dakota, is not without a certain affection and respect for the little wolves he has hunted so relentlessly. He raised one for a year and a half, collecting its urine to use in coyote sets. "In the evenings when I came home," he said quietly, "he'd leap all over me."

His regard for the coyote is shared by others who know the animal best.

Indeed, the coyote may be the most remarkable wild creature anywhere in North America. Other wildlife—wolves, grizzly bears, bison, pronghorns and prairie grouse—have vanished, or nearly so, in the wake of plows, fences, fires, and guns. But the adaptable coyote, still unprotected by law almost everywhere, victim of persecution by stockmen and their agents for decades, has not only survived but prospered. Today, the coyote controversy, with sheepmen demanding that losses from coyotes be reduced by any means, is raging louder than ever.

How has the coyote managed to cope with civilization? It possesses an uncanny sensitivity about threats to its welfare. With an inherent intelligence and catholicity of food habits, the little song-dog of the prairies has gone tiptoeing over the landscape, exploring new places, sampling new foods, testing new hazards, to become the central figure in the most compelling wildlife success story in the history of our continent.

Amazingly, the coyote has expanded its territory southward into Central America, northward seven thousand miles to the Arctic coast of Alaska, and eastward to the Atlantic seaboard. Today few states lack coyote populations. Furthermore, the animals have moved steadily closer to cities (or cities closer to coyotes). Residents of Toronto are no longer surprised when they hear the coyote's song. No one knows the size of the coyote population in Ontario's metropolitan York County, but the little wolves are seen, heard, and shot often enough to establish them as a permanent part of the fauna.

A female coyote cohabiting the outskirts of Philadelphia with a local dog raised a litter of pups. Citizens, fearful of rabies, methodically hunted them down and killed them in the suburbs of the City of Brotherly Love. Another female coyote lives in Denver's Fort Logan Cemetery. Her mate, a Labrador retriever, annually bequeaths her a litter of pups with bodies like coyotes but fur of varied quality and color. A coyote recently attracted attention in Fort Collins, Colorado, by hunting across the yard of the courthouse.

Probably no city has a more pronounced coyote problem than 464-square-mile Los Angeles, where song-dogs howl nightly from the Hollywood hills and on occasion sip chlorinated water from the world's most costly swimming pools. They also perform such antisocial acts as frightening people and killing dogs and cats. Two thousand coyotes are said to live within the Los Angeles city limits.

Yet the coyote still belongs to open country where ranchers, farmers, and campers spot it in the buffalo grass and sagebrush. The male stands twenty-one inches high at the shoulders, measures forty-eight inches from nose to tip of tail, and averages twenty-five pounds. The female is somewhat smaller. A bit on the shaggy side, the coyote has pointed, erect ears and a bushy, drooping tail. It is mostly gray with a tinge of yellow on flanks and neck, and black on forefeet, tip of tail, and at the tail's base.

This wild canine is heard more often than it is seen. In early evenings on a sagebrush hill, it lifts its head to the yellow half-moon and opens a concert with a series of sharp yips and yaps which soon stretch into a long and doleful howl. Almost at once there is an answer from a nearby hill. Then a third coyote speaks up, and then more, until it is difficult to tell how many animals have joined in the songfest, all coming in on cue and filling the evening air with an unforgettable medley. It is a haunting reminder of the time when westbound settlers listened tremulously to prairie wolves howling beyond the light of their campfires.

What does the singing mean? Are the coyotes barking at the moon, talking to each other, or merely enjoying themselves? At Colorado State University animal behaviorist Philip N. Lehner and a dozen other scientists have worked on this question in the course of coyote research. "Coyotes undoubtedly vocalize a great deal of information to each other," Dr. Lehner says. Franz Camenzind, one of his students, has identified four kinds of long-range vocalizations. First there is the "lone howl" of a single animal, which may serve to bring the group together. Second there is the "lone bark-howl," which may be a warning or challenge between individuals of different social groups. The third call Camenzind has identified is the "lone bark," considered a danger warning. The fourth is "yipping-howling," a group activity which the researchers believe reinforces group cohesion and maintains territorial integrity as surely as the songs of nesting warblers and wood thrushes. And there may be still more calls with other meanings.

Strangely, coyotes may be set to howling by sudden loud noises. Young coyotes living in laboratory cages at Colorado State University set up a chorus of howls whenever trains whistle at a crossing a half-mile away. Folklorist J. Frank Dobie reported that long ago, on the San Saba River in Texas, a hermit of Scottish descent brought out his bagpipe on fine evenings to render ancient ballads to the desert moon. As he played, coyotes round about tuned up and harmonized.

Another coyote that can be induced to howl on cue is the young female owned by the family of F. Robert Henderson, extension specialist in wildlife damage control at Kansas State University at Manhattan and an authority on coyotes. Taken from the den with her eyes still sealed, the tiny furry pup was tediously hand-fed with bottle and nipple by the Henderson family. A coyote raised by other coyotes will be forever wild and fearful of people, perhaps sensing that perpetual caution is the price of life and that a single mistaken judgment or overlooked clue is the prelude to tragedy. But the Hendersons' coyote remembers no coyote parents, no dark cavern in the hillside, no melodious ballads sung of an evening under wild prairie skies. At the sight of her owner, she bounds joyfully about her pen. When turned loose she stays with the people who have reared her, romps with the children, and responds to the name they gave her—Rattlesnake.

One afternoon the Henderson children accompanied us into their yard and with high-pitched voices began mimicking the song of the coyote. Rattlesnake lifted her nose to the sky and harmonized in solemn concert. Anytime the Hendersons want to stage a sing-along, Rattlesnake obliges.

Later we took Rattlesnake into an open field where she was turned free to exercise and pounce on grasshoppers. A passing car slowed perceptibly so three young men could watch the animal. "No coyote is safe," Henderson commented, and recalled the case of the coyote that was a village pet: Nobody bothered it until a stranger saw it playing with dogs at the edge of town, skidded to a halt, and shot it. Another pet coyote was shot from the road while sitting on its family's front porch.

As evidence of the coyote's high level of intelligence, field observers have often pointed out its hunting techniques. While one coyote chases a rabbit, another may wait behind a bush to leap upon the prey as it runs by. Bill Pullins tells of sitting on a hillside with binoculars and watching a family group of coyotes on a jackrabbit drive. The ten coyotes were lined up in a straight row, and "when one flushed a rabbit, they were quick to get it. Quite a sight." Frank Dobie described a coyote that leapt stiff-legged into the air and otherwise clowned around for the benefit of a spellbound jackrabbit as another coyote crept up and surprised the rabbit from behind. Some scientists speculate that coyotes have ultrasonic hearing which enables them to locate mice beneath snow when there is no exterior sign of movement or sound audible to the human ear.

Retired trapper Lloyd W. Hutchinson of Lovell, Wyoming, who was a government trapper for eight years and then supervisor of all federal trappers in the state, is no enemy of the coyote, either. He looks back on his work as a job that needed doing, and was not displeased when "the ecologists stopped us from using poisons." Through the years he became convinced that "we have created a race of supercoyotes." He believes coyote dens are harder to find now. In places where some coyotes were killed by eating from poisoned carcasses, other coyotes no longer will eat carrion. Coyotes that once sat on a hill and studied humans entering their territory now run and hide. "They're the smartest wildlife we have," Hutchinson says. "I would not want to see them eliminated."

"After watching many coyotes, I really believe they have a sense of humor," wrote mammalogist Victor H. Cahalane. According to a story told by Dobie, it would certainly seem so. A ranchboy who used to drive a horse and wagon to town regularly saw a coyote looking on at a particular point along the way. One day the boy's fat old shepherd dog leaped from the wagon to chase the coyote. But instead of coursing off across the prairie, the coyote began to run around the wagon in a wide circle. It gained on the dog so fast that after a few laps, the shepherd was running full-speed to keep ahead. Finally the dog cut across the circle and jumped back into the wagon. It cringed there in disgrace while the coyote sat down beside the road and watched the wagon move out of sight.

Although not strictly monogamous in the breeding season, if left undisturbed coyotes will maintain their family ties. Male and female stay together until one of them dies. They may hunt together, and both will help provide for the young. When the female is in the den with her newborn pups, the male delivers food to her. (For some unknown reason the male at this time may kill many more animals than the coyotes need for food.)

Coyotes usually mate in midwinter, and the female carries her young for sixty to sixty-three days. By early spring she has prepared several dens, one of which will become the nursery. The pups are born by mid-April in Northern states and somewhat later in the South. There may be five to seven of them, with eyes sealed and bodies covered by dark gray woolly fur. After a few days the female moves outdoors, where she spends hours resting downwind from the den but is alert for signs of danger. If the family is threatened, the parents may move the pups. Joe Van Wormer wrote in *The World of the Coyote* about an adult that moved four pups a distance of five miles—thirty-five miles of travel—in a single night.

By the time the pups are three weeks old they may totter up toward the mouth of the den to stand blinking in the strange brightness of the spring sun. Their wobbly legs gain strength rapidly, and in the following weeks they spend much time wrestling, tumbling, and pouncing on insects around the den's entrance.

Many coyotes perish before they are one year old as victims of traps, highways, and hunters. A large percentage of those killed by predator-control agents are young—newborn pups either dug from dens or dragged out by a long piece of barbed wire twisted into their fur. The ones that survive may stay with the family group or disperse into new territories, depending on food supplies and social pressures. The first winter may be the most hazardous time of their lives.

How far does a hunting coyote travel? According to Donald S. Balser, U.S. Fish and Wildlife Service biologist at the Denver Wildlife Research Center, the animal probably spends much of the year within a territory covering only a few miles. Males in established territories may cover at most twelve square miles, and females considerably less, perhaps four or five. Young coyotes seeking new territories, however, may travel much farther.

The coyote is the constant enemy of such smaller animals as rabbits, mice, and prairie dogs. But larger animals may be a serious threat to the coyote. A state conservation officer in Ely, Minnesota, saw wolves catch a coyote out on a lake's winter ice and tear it apart. And Lloyd Hutchinson often witnessed eagles diving on coyotes repeatedly. Eagles have been known to kill coyotes.

Occasionally coyotes attack and kill very large animals—adult pronghorns, deer, and elk—but the tables may turn and the coyote may perish beneath sharp hooves. In Wyoming a buck deer was seen running down a coyote and battering it so badly that the coyote barely managed to escape. When Franz Camenzind was conducting wildlife behavior studies on the National Elk Refuge in Jackson Hole, Wyoming, he witnessed the complete demoralization of a coyote pursued by a female antelope to the point of exhaustion. The coyote escaped only by running into deep forest. Pronghorns, whose fawns often fall prey to coyotes, have been known to trample their natural enemy to death.

Strangest of all the coyote's relationships is that with the badger. Ever since the times of ancient Indians, men have seen these two wild predators traveling together and tumbling and playing together. "The coyote on his long legs can travel a lot faster than the badger," says trapper Lloyd Hutchinson, "but pretty soon he stops and turns around and waits for the badger to catch up." There is an obvious advantage for the coyote in the association: The badger, with its long powerful claws, digs out burrowing rodents, and any that escape are pounced on by the waiting coyote. Yet there is no record of the coyote sharing its catch with its short-legged companion, and what the badger gains from the partnership is not clear.

The arrival of European man brought a new and fearsome dimension into the coyote's world. Coyote-haters have been known to cut off the lower jaw of a trapped coyote and then release the animal, or wire the jaws shut and turn the coyote loose to starve. Weekend trappers may leave a coyote in a trap for days until they return and mercifully kill it with club or gun or by stomping to avoid damage to the fur. Twisting, turning, and gnawing, the frantic animal may amputate a foot and escape before the hunters' return. Peg-leg coyotes are numerous in the West; hampered in their hunting, they often survive by learning to catch sheep.

Not long ago the people of Millinocket, Maine, had their first look at a coyote when a fifty-five-pound male was shot and hung up on public view. *Maine Times* editor John Cole later wrote in *Harper's Magazine*, "Townspeople gathered around like primitive villagers. They cursed, reviled, and spat upon the dead coyote."

Man's everlasting efforts to kill coyotes had their genesis in his age-old unreasoned hatred for wolves. Sometimes entire armies of hunters have set off in righteous pursuit. In 1905 David E. Lantz of the U.S. Biological Survey, now the Fish and Wildlife Service, wrote about witnessing the annual "wolf hunt" on Thanksgiving morning near the village of Chattanooga, Oklahoma. Spectators from miles around arrived by farm wagon to watch, in holiday spirit, as one hundred fifty men and one hundred twenty hounds began their drive. "The sport," Lantz wrote, "was fast and furious for a time, but when a little later the dead and captured wolves were brought together in the town, they were found to number only eleven in all. Two of them were roped by cowboys during the drive and killed with pistols. Two were dragged to death at the end of lariats. Seven were caught by the dogs in the roundup, and two of these were brought in alive. Such hunts . . . afford an agreeable break in the monotony of frontier life. Their purpose, however, is never admitted to be that of sport, but to kill coyotes."

Writer Bob Parkhurst of Tulsa, Oklahoma, recently recalled such a hunt in 1941 in northeast Kansas. Eight thousand boys and men (many of them soldiers from Fort Riley) who were directed by airplanes equipped with loud speakers killed two hundred coyotes in two counties. A similar hunt was held on New Year's Day 1974 at Wheaton, Kansas, for the benefit of the Heart Fund.

Today, in the Dodge City area of southwest Kansas, men on motorcycles play a weekend game. A dozen Honda-mounted riders line up across the prairie with their motors roaring. Eventually a coyote panics and dashes out of cover ahead of the line. Machines speed forward in formation to pursue the animal to the point of exhaustion. Then one rider executes it with a .22 handgun. After a brief rest and a round of congratulations, the motorcycles re-form their line and snort off in search of other victims.

Bounties have been offered for killing wolves since the eighth century B.C. in ancient Greece. Although the bounty system never controlled animal populations, it moved to the New World along with the white man. On American frontiers some of the earliest community gatherings were "wolf meetings." A Massachusetts law dated 1648 offered any Englishman a reward of 30 shillings for the head of a wolf; Indians were offered 20 shillings.

The bounty system moved west to the Great Plains and the Rocky Mountains. During the 1800s professional wolfers hired out to ranchers to trap, poison, and shoot wolves—and coyotes. Old ways die slowly: In Michigan, legislators are still debating the wisdom of halting bounty payments on coyotes. The State of Maine recently defeated a hard-fought campaign to establish a $50 bounty on its four hundred coyotes in the name of "saving the deer."

Predator-control efforts intensified as stockmen across the West became better organized. In 1907 a record kill was reported for the national forests—

1,800 wolves and 23,000 coyotes. Meanwhile, ranchers were exerting pressure to get the federal government into the predator-killing business. In 1914 Congress authorized an expenditure of $125,000 for the control of wolves, and the die was cast. Sheepmen then wanted the program extended to the killing of coyotes and quickly won their point. By mid-1916 much of the western range country had been efficiently organized into predator-control districts; federal trappers and supervisors were assigned to them; and a virile new government agency—the Biological Survey's Branch of Predator and Rodent Control—had made the old-time free-lance bounty hunters obsolete. Within thirty years the federal program would claim 1,792,915 coyotes.

In 1939 when the Biological Survey was moved from the Department of Agriculture to the Department of the Interior, its Branch of Predator and Rodent Control went along. It prospered, and was reorganized in 1965 as the Division of Wildlife Services, devoted to wildlife enhancement and pesticides monitoring and surveillance, as well as predator control. (It was recently split into the Division of Technical Assistance and an animal-damage control unit.) Wildlife Services has spent about $8 million annually, including cooperative funds from state governments and private organizations, for predator control. In 1971, for example, $8.1 million was spent for killing 121,135 animals—75,661 coyotes, 6,608 bobcats, 234 bears, 80 mountain lions, and 15 additional species ranging from gray wolves to weasels.

Wool-growers, claiming they are victimized by coyotes and insensitive politicians, have been the major clients of the Division of Wildlife Services. Wherever western sheepmen assemble, talk quickly turns to coyotes, and other enemies of sheep are minimized. State conservation officer Michael Muck of Philip, South Dakota, described how three hundred head of sheep in his area died last year from eating moldy corn. "That's just one of those things," he said. "But let the ranchers lose one sheep to a coyote, and they damn near have a cardiac arrest."

I had been told that if I talked with rancher Otto Wolff, who raises sheep in the Black Hills and Badlands of South Dakota, I would learn all I needed to know about the feelings of many western sheepmen toward coyotes. Wolff, a tall, salty, outspoken enemy of predators in general and coyotes in particular, said, "There is no such thing as good predators and bad predators. All of them are predators. Either you have a blanket program or you have no program. The best tool we ever had was poison, and the best poison was ten-eighty. These misguided people had the gall to deprive us of one of our tools of management. We're not telling people in the East how to control their riots and run welfare, and we don't think they should tell us how to run our business.

"I have been running sheep for forty-seven years," Wolff continued. "By running sheep on public land, we are really making use of marginal land which is not good for one other damn thing. This is in seventeen western states, and that's a lot of fiber and meat for the country."

Then he zeroed in on the coyote. "Without the coyotes we would have damn few problems. As an industry we have made excellent progress with vaccines and dips. We just don't lose lambs anymore. Oh, maybe one once in a while from blowfly after a lamb gets stuck on a stick. These people that want to save the coyotes want to preserve a bunch of bastards that don't do us any good at all.

Balance of nature? To hell with that. We operate on a slim margin. Losses to predators are right off the top. There are millions of acres where we could let these wild animals run, but I don't want them on my land." ("My land" includes national forests where Wolff has grazing privileges.)

"The coyotes don't just take the sick and the old," Wolff argued. "The old and sick ones are left alone. The coyote wants only the running sheep. They like the strong young lambs." I asked him how the ranchers know the number of lambs lost to predators. "We estimate them, of course. You have to. You know how many sheep you put out on range, and you know how many you bring back in. If any die you see buzzards and eagles, and you go find them."

Wolff belongs to a sheepman's association that has its own predator-control plane and also hires other fliers. "One guy got seven coyotes in one hour shooting from a helicopter," he explained, "and we paid him twenty dollars for each of them. Then the fish-and-game people raised hell and made us stop. The trouble with the fish-and-game people is they listen to those damn biologists." The ideal solution, he believes, would be to move predator control from the Department of the Interior to the Department of Agriculture's Extension Service, because "Those people are our friends."

Cattlemen, however, often have a different view of the coyote. Although they may report occasional losses of calves, and such losses are said by some to be increasing, it is still easy to find cattle ranchers who consider the coyote a friend. "I've seen them at their worst and at their best," said rancher–writer Dayton O. Hyde of Chiloquin, Oregon, "and I'd ride a long way to stick up for them." Another cattle rancher declared, "We know coyotes kill sheep, but not fast enough."

Cattlemen favor the coyote because it eats the rodents and rabbits that eat the grass their cows need. Some years ago H. T. Gier of the Kansas Agricultural Experiment Station calculated that one coyote saved $12 worth of grass which would otherwise be consumed by rabbits and rodents. And Dayton Hyde, testifying before a Senate committee in 1973, placed the savings per coyote at $88, more than double the price of a lamb at the time (40 cents per pound live weight).

"We need the coyote," says South Dakota rancher Skee Rasmussen, a former member of the state's Department of Game, Fish and Parks. In season he and his friends hunt many kinds of game on his ranch, but he draws the line firmly at coyote hunting because "they have a hard time." He even maintains a sizable prairie-dog town on his property—"the coyotes need them."

Some years ago when the poison 1080 first threatened the coyote, cattle-man Charles Price of Philip, South Dakota, became so concerned about the little wolves' future that he located a zoo with surplus coyotes and purchased a pair of eight-month-old pups. Raised in a pen on his ranch, they produced three litters of six or seven pups each. The gate of their pen was kept open so that the coyotes, young and old, came and went at will until eventually all of them had scattered over the ranchland, rebuilding the wild-coyote population.

But in the main, for a century and a half hunters have pursued coyotes, and ranchers have condemned them. Professional wolfers and federal and state agencies have bountied, poisoned, trapped, and shot them, and dug their pups from dens. The individuals who saw in the coyote the marvel of a highly adaptive organism, or who savored its evening concert from the hills, rarely defended the little wolf, assuming that, for all its charming mannerisms, it was in reality the sheep-eater which sheepmen claimed. There was little scientific effort to unearth the facts. Reviled, detested, and condemned, the coyote was attacked continually.

A few years ago, however, conservationists became concerned about the many nontarget birds and mammals falling victim to coyote poisons, and demanded that the environment be spared doses of thallium sulfate, 1080, cyanide, and strychnine. Finally in 1972 President Nixon issued an executive order banning the use of these poisons on public lands. The order was followed by Environmental Protection Agency action making it illegal to ship such poisons across state borders, and by pesticide legislation that made the use of poisons in predator control illegal.

The ban on poisons immediately brought forth claims of coyote population explosions from stockmen. Many biologists, however, believe coyote populations are subject to cycles controlled by little-understood natural factors. "Coyotes were already on the increase before the order was issued," says Frank Martin, former manager of Montana's Charles M. Russell National Wildlife Range.

Spokesmen for the sheep industry descended on Washington to plead with legislators and testify before committees, painting a picture of an ancient, honorable industry doomed by the coyote. The sheepmen wanted Congress to pass a new law that would relax the ban on poisons and give federal funds to states to increase their predator-killing campaigns. State agriculture departments rather than wildlife agencies would administer these programs. As a result of the pressure, the Environmental Protection Agency has approved use of the M-44 spring-powered set gun, a device which shoots a cyanide capsule into the mouth of any animal tugging at a baited wick, whether coyote, fox, or domestic dog. Officials say the cyanide gun will be less hazardous to humans than its predecessor, the explosive-type "coyote-getter" widely used for twenty-five years before poisons were banned.

Demands for legislation aside, the anti-coyote clamor also resulted in a massive, long-overdue research effort into the secret world of the coyote. After spending millions of dollars on coyote extermination, man very belatedly began to learn something about the animal itself.

The little research accomplished earlier had invariably concentrated on the coyote's food habits. Pioneering scientists of the U.S. Biological Survey had probed the contents of fifteen thousand coyote stomachs and demonstrated that coyotes, testing for palatability, will eat almost anything. (This was no news to cowboys, who had learned long before not to tie their horses with rawhide lines or coyotes would free them by chewing on the leather.)

Although largely carnivorous, the coyote is known to eat berries, peaches, pears, apples, plums, apricots, and mesquite beans. Some coyotes are so fond of watermelon that they constitute a major economic threat to melon farmers in the irrigated valleys of the West. They are said to choose only the ripe fruit, and take only a bite or two from each one. Other coyotes, especially pups in late summer, will feast on grasshoppers until their stomachs become distended. They eat mice, chickens, turkeys, wild birds (if they can catch them), and fish. They have been known to consume pine needles and even rubber from an old hot-water bottle; evidently, they have no prejudices against anything that is or might be edible. The coyote is an opportunist—a supreme example of making do with whatever is available, including the flesh of its own kind.

Nevertheless, a 1971 predator-control report by a committee of noted biologists pointed out, "No coyote food-habits study has ever shown livestock to be a major part of the diet." In fact, studies have revealed repeatedly that rabbits and rodents account for the largest part of the coyote's natural diet with carrion ranking second.

Today, for the first time, coyote research is advancing beyond the stomach-probing of yesterday. Government bureaus, colleges, universities, and private agencies are pursuing the song-dog on so many fronts that no one knows for certain how many studies are under way, but a preliminary list compiled in 1973 identified ninety-two coyote-related research projects.

In his office at the Denver Wildlife Research Center of the U.S. Fish and Wildlife Service, biologist Donald S. Balser explained, "We are investigating the predator-livestock problem from three angles: damage assessment, depredations control, and predator ecology." Hundreds of sheep and coyotes are wearing miniature radio transmitters and being tracked electronically by field agents. In three hundred twenty-eight locations in seventeen western states, a newly devised system uses special scent posts on fifteen-mile-long lines to measure coyote density.

Other research seeks information on productivity, parasites, sex ratios, population age structures, causes of natural mortality, dispersal of young, daily movements, visual acuity, vocalizations, and relationship to deer, antelope, wolves, foxes, sheep, and mice. There is talk of chemiosterilants to curb coyote populations. Balser's co-worker, Samuel B. Linhart, is investigating the possibility of discouraging coyotes from eating sheep through the use of electrical shock, chemicals, and special fences; researchers at South Dakota State University are pursuing the same possibility through the use of ultrasonic sounds.

Studies in Texas and Louisiana aim to rescue the endangered red wolf. It is threatened, at least partially, by hybridization with coyotes. Smaller than the gray wolf, it crosses readily with the somewhat smaller coyote which has invaded its range.

For its part in the research effort, the U.S. Department of Agriculture made available to land-grant colleges funds for finding ways to frighten, disturb, or disgust coyotes enough to discourage them from killing sheep. Texas A&M University received $7,000 for research on coyote-proof fences; South Dakota State University was given a like amount to study "acoustical repellents"; the University of Wyoming got $30,000 to find "gustatory repellents" and Colorado State University received $30,000 to begin a study of "olfactory repellents."

One study of olfactory repellents is being conducted thirty miles north of Fort Collins, Colorado, at the Maxwell Ranch. Captive coyotes have been trained to run across an enclosure to pick up a piece of meat in the midst of strange and, hopefully, discouraging odors issuing from a buried pipe, in an effort to determine whether sheep can be treated with agents that will repel coyotes. Results have been promising.

Near Laramie, Wyoming, caged coyotes are being used in a University of

Wyoming search for substances that repel coyotes because of their taste. Wildlife biologist Douglas M. Crowe explains they have found that commercial dog repellent, oil of citronella, and Cayenne pepper turn the animals away. Further testing will involve placing repellents on collars worn by sheep to learn whether coyotes will avoid them.

"Coyotes," says Bob Henderson at Kansas State University, "become specialists. They take what is easiest to obtain." They may become specialists in hunting rabbits, mice, watermelons, or prairie dogs. They know their hunting territories and learn where they can most easily obtain their customary foods. If wild foods become scarce, the resourceful coyote may turn to domestic livestock, and if successful, livestock will become its new specialty. Then the coyote is conditioned to eat sheep for the rest of its life.

Like the coyote, Henderson is a specialist. He concentrates on coyote damage control—but not the species-wide type of eradication so long practiced by federal trappers. He believes that only the trouble-causing coyote should be removed and that ranchers and farmers can perform their own predator-control work. To this end he travels about Kansas encouraging farmers to consider the biological aspects of coyote control, explaining why a coyote kills in the first place, and teaching preventive measures. Following the Henderson methods, a rancher can soon hang up his traps and go on about his other work.

"Our entire program," says Henderson, "is based on the conviction that instead of coyote control we should be practicing coyote *damage* control. We're not trying to kill off all coyotes. We know that can't be done, and we're not at all certain it would be desirable. What we want to do is remove animals known to be destroying stock. At times a few coyotes can become a very serious problem. Each case should be investigated separately."

Franz Camenzind's studies at Jackson Hole support this approach to coyote problems. "Coyotes are territorial," he told a senate hearing, "and resident coyote will act to keep neighboring coyotes from encroaching upon their area. If depredations are occuring, it is probably the result of only a few local animals."

Massive coyote-control programs over vast areas of the West, he concluded, resulted in an "extraordinarily wasteful overkill."

For Interior's Division of Wildlife Services, Kansas has been a sore spot within a solid block of states where federal animal-control people are active. To the north, Nebraska utilizes federal trappers in fourteen counties, and the cost of their predator-control work in 1973 was $126,000. To the south, Oklahoma also uses federal agents, and the cost of their work in 1973 was $289,000. But in the same year Kansas, with one of the country's highest coyote populations, an important livestock industry, and no federal trappers at all, spent only $21,000. Yet Henderson found that forty-four of the one hundred and twenty farmers he worked with in 1972 stopped their sheep losses completely by doing their own coyote trapping, and the rest at least reduced their losses. Admittedly the extension trapper plan might be less effective on open range; but it is already being practiced in Missouri, Arkansas, Iowa, eastern South Dakota, and the Canadian province of Alberta.

As for the effectiveness of massive coyote-killing programs, there is evidence that coyotes possess an inherent capacity to compensate for unusual population losses. Biologist Frederick F. Knowlton, writing in the *Journal of Wildlife Management* in 1972, disclosed that the average size of a coyote litter varies according to current population control factors. In south Texas, where coyotes were extremely abundant, he found that coyote litters averaged 4.3 young. But in the Uvalde section of south Texas, where "coyote numbers [were] drastically reduced by intensive control efforts, the average litter size was 6.9." Furthermore, in areas where coyotes suffer depressed population levels, females appear to begin breeding at younger ages than elsewhere.

Similarly, Colorado State University researcher Franz Camenzind found that coyotes in Jackson Hole, Wyoming, where the population is free from traps, gunning, and poisons, have litters averaging 4.5, as contrasted with six or seven pups in areas where heavy control is practiced. Camenzind concluded, "The more coyotes that are removed from the area, the more pups the remaining adults will produce."

Camenzind believes that coyote populations probably follow cyclic patterns when they are left to such natural forces as disease and food shortages. He speculates that control efforts might have interfered with population fluctuations in the past and kept the natural controls from working. "Undoubtedly, the control programs kept coyote populations from reaching maximum numbers and crashing naturally. It is conceivable that the total number of coyotes, and thus the amount of predation, was actually greater than if the coyotes had been allowed to cycle normally."

For a half-century the wool-growing industry has been faltering. The number of sheep in Kansas, for instance, slipped from 700,000 in 1943 to 325,000 in 1973, and the story has been the same in other states. The decline has occurred despite massive federal subsidies in the form of predator control, low grazing fees on federal land, and price supports on wool and mohair. Flocks once tended by shepherds now roam the Western mountains alone because of the high price of labor and the changing work ethic. Land values rise, taxes go up, feed prices increase, margins of profit fall. And the coyote gets the blame.

Just when it began to appear that the war between coyotes and sheepmen would continue forever, I met sheep expert Hudson Glimp. Dr. Glimp is a former research specialist at the University of Kentucky and the U.S. Department of Agriculture, and a former faculty member at Oklahoma State University. He lives in Wheatland, Wyoming, where he manages a budding program that has sheepmen around the world watching with interest. He wants to revolutionize the sheep industry and may well do so in a manner favorable to the beleaguered coyote. Glimp is general manager of the new $6 million Y.O. Ranch installation, which will have a "confinement-rearing system" for sheep. Plans call for the housing of five thousand ewes; lambs will be raised indoors on a production-line basis and fed artificial milk made of soybeans. The world's most modern sheep-shearing plant will be located nearby; also a new slaughterhouse from which eight hundred lambs a day will be packaged for the retail market. A predator-proof fence will enclose the entire operation. "We will eliminate the two biggest threats to sheep, which are parasites and predators," Dr. Glimp declared. He believes that children will have to go to museums to see sheep unless American wool-growers move in this direction.

One wool-grower who has already done so is George Ahlschwede of Manhattan, Kansas, a sheep specialist at Kansas State University. Dr. Ahlschwede's ewes are confined at night; his lambs are fed inside buildings with slatted floors. In four years he has lost no sheep to coyotes, although coyote tracks are seen frequently nearby and the Manhattan area has one of the country's largest coyote populations—five per square mile.

Perhaps with the advent of new sheep-rearing methods, the versatile song-dog has won another round. And perhaps the ancient Indian tribes who believed that "brother coyote" would be the last animal on Earth were truly prophetic.

THE UNCANNY CAT, BY RONALD ROOD *The bobcat is a more frequent resident of America's wild and semi-wild places than its rare encounters with people would suggest. And perhaps it is best that those encounters are few. For whether you merely hear the frightening night "music" of the bobcat's courtship, or chance upon one face-to-face in a deep swamp, the experience can only be described as unforgettable.*

The water was creeping above my ankles, but I didn't dare look down. Even though uncornered wildcats aren't supposed to attack people, this one could be different. Surprised by a boy only ten feet away, he might well figure he was cornered.

I'd been stepping from one hummock to another in a swamp, looking for turtles and frogs. As I put my weight on each hummock, it would sink. So my progress had been marked by a strange ballet as I jockeyed my way toward a little grassy island twice the size of a kitchen table. I was just two leaps away when I raised my head for a last check of its position. It was then that I was stopped by those blazing eyes.

His mouth opened in a soundless snarl. Those four white fangs looked long and terribly sharp. Brown and white streaks gave his face a look of savagery as he laid his ears back against his head. The yellow eyes narrowed until the pupils were just slits. And there we both stood.

Then came the uncanny part. While my heart pounded and his face etched itself on my mind, I suddenly discovered he was gone. He hadn't bounded away or plunged into the water. He had just faded from sight. Those eyes had melted into the background while they held me spellbound. The open mouth somehow became a dry leaf. Those facial streaks turned into the brown grass of the April swamp. Only the gentle swaying of the vegetation told me I hadn't dreamed the whole affair.

Whether he finally swam away or still lurked, hidden, I do not know. Nor did I try to find out. That was my first meeting with a wildcat. It lasted only about ten seconds. But they were a long, indelible ten seconds. Since then I've talked with other people who have met wildcats, and nearly every one of them reports the same reaction—a steady, unblinking stare and an unhurried, ghostlike retreat. They all use the same word to describe the encounter: Unforgettable!

What are the chances of the average person seeing a wildcat? Perhaps this question should be turned around: What are the odds that a camouflaged wildcat might peek out from its swamp or thicket at *you?*

The answer may be a bit surprising. In the Southwestern scrub, you may come under the gaze of a wildcat as often as once a week during a camping jaunt. A rancher friend of mine hunts wildcats and gets about one a week, while in my own Vermont woodlands, one very successful hunter bagged an even dozen in twice that many weeks.

Apparently, if one is looking at you—whether you're aware of it or not—it's only curiosity. A neighbor of mine had the unnerving experience of being followed by a wildcat for more than a mile as he walked through the New Hampshire woods with his little dog. The dog bristled and growled and clung so closely that his master finally picked him up and carried him. Occasionally they saw the wildcat peacefully trotting along behind them, apparently having no harmful intentions whatsoever.

Barring such an episode, the opportunity of glimpsing a wildcat in the eastern woodlands may occur perhaps once in ten or fifteen years. In the more open South and West, says my rancher friend, the average person out in the bush may see a wildcat once a year.

The creature has such a grip on the popular imagination, however, that you're probably not far from other "wildcats" even as you read this. You may have a Wildcat parked in your garage or resting on your boat trailer. Marine Navy pilots flew Wildcats at the enemy in World War II. There are Wildcat chain saws, Wildcat motorcycles, and—intriguing as the prospect sounds—a Wildcat Sewer Service in a town not far from my home.

If you've driven through the Catskill Mountains, you've been in wildcat country, too, in more ways than one. Henry Hudson dubbed the area's rushing river *Kaats Kill*—Dutch for "Wildcat Creek." My wife Peg and I once stayed at the Wildcat Motel in one of the region's little towns. It was a singularly uneventful visit.

There are wildcat (take-a-chance) oil wells, wildcat stocks, wildcat enterprises of many kinds. Every one of our wars since the Revolution has engaged the services of Wildcat regiments, squadrons, or other fighting units. And doubtless nearly every state has its fighting Wildcat high school and college athletic teams, even in areas which may not be blessed by the presence of the real animal.

At first glance, a wildcat looks ordinary enough. It's about half again as large as a good-size tomcat. But as you look closer, you realize that's where the resemblance stops. Averaging perhaps twenty pounds, the wildcat seems to be a miniature cross between a tiger and a leopard, with a bit of mountain lion thrown in. Its rust-brown coat shows spots and flecks above and a suggestion of dark stripes below, blending into a white belly. Heavy lines on its wide-flaring cheek fur break up the outline of its face so it can see without being seen. A little tuft of hair on each ear serves as an antenna which is sensitive to sounds or air currents. Its whiskers, bedded in delicate nerves, may lie back—or reach out to determine if a certain opening will admit its body.

There's nothing quite like a wildcat in action. While it's ordinarily so peace-loving that even a Chihuahua could probably put it to flight, once aroused it can whip nearly any dog alive. A flick of powerful muscles, and those velvet paws suddenly bristle with eighteen knives—four claws on each hind foot and five in front. Spitting defiance, it offers the enemy one last chance to retreat. The ears and sensitive whiskers are laid so flat against the head that they seem to disappear. Then it springs to the attack, clinging with front claws, roweling and digging with the rear ones, biting deep with its fangs.

Indeed, a wildcat sometimes hates to quit. Cornered by more than one enemy, it leaps from one to the other in such a blaze of claws and teeth and fur that its bewildered tormentors may soon become the tormented. After seeing one wildcat in action, I know now what the pioneers meant when they said something was "as easy as letting go of a wildcat."

But there's the lighter side. As with other cats, its tail twitches when the owner is excited. Somehow, though, the wildcat's little six-inch appendage merely looks ridiculous as it flips back and forth. It's this afterthought of nature which has earned it a second name—bobcat.

It is also called bobcat for another reason. A running wildcat bobs up and down like a rabbit. Its relatively long hind legs give it a gait better suited to a short dash than a long run. Hunters know this, and count on a bobcat to take refuge in a tree after their hounds are fast on the trail. Otherwise it might go hard for the dogs, indeed.

The Canada lynx of the North is the only other American cat resembling the bobcat. Its average size is slightly larger (up to thirty-five pounds), though some bobcats are fully as large. Stub-tailed like its southern relative, the lynx has longer eartufts and a more uniform brownish coat. Its great padded feet grow "snowshoes" of stiff fur in winter. Of course, the bobcat would scarcely need snowshoes at the southern limit of its range, in central Mexico—nor, for that matter, over most of its territory. This covers much of the continental United States, the bobcat ranging north to overlap the lynx near the Canadian border.

Scientists, in distinguishing between the two, named the bobcat *Lynx rufus* —which, naturally, means "the rufus-colored lynx," although individuals of Southwestern chaparral and canyons are nearly as pale as desert sand. When biologists came to the Canadian lynx, however, they neatly sidestepped a description of the gray-brown animal by calling it *Lynx canadensis*. Otherwise the two animals are much the same, although the Canada lynx is more retiring and less able to abide the presence of man.

The modern bobcat is the result of a long period of adjustment in nature. Over the eons the cat family evolved from some remote Old World ancestor: tigers and lions to catch zebra and antelope, mountain lions to catch deer, the Canada lynx to prey on the snowshoe hare, little Asiatic and African cats to catch birds and mice and give rise to our household felines. Somewhere, perhaps eight million years ago, the forerunners of the bobcat began to develop a taste for the forerunners of our rats and rabbits. They've been together ever since.

Even though today's bobcat can take the presence of man in small doses— setting up residence in overgrown brushland or making nightly rounds along gravel roads—you'll be lucky to see even one living specimen in the wild. Its eyes and ears are easily keen enough to warn of your coming, though its nose is less sensitive. And what I learned as a boy apparently still holds: No wildcat in its right mind would, unprovoked, attack a human being.

Once in a while, however, one makes a mistake. Stanley P. Young, who has known these animals for years, tells in his fascinating book *The Bobcat of North America* of a hunter who went out after coyotes in Texas. Settling down beside a bush, the hunter waited for some potential target to show itself. To stack the odds in his favor, he decided to call a few times in an imitation of a coyote.

He got more results than he'd planned. With a yowl, something hit him in the back. It was a full-grown bobcat. Claws out, teeth bared, she lit into him until she discovered her error. Then she bounded away.

Shaken, the hunter tried to piece together what had happened. After a search he found a den of bobcat kittens nearby. She had been protecting her family from this invading enemy—something she would not have done if she'd realized the "coyote" was a human. She'd merely have slunk away, and he wouldn't have been the wiser.

Though bobcats are hard to see, they're nowhere near as hard to hear. They use the same language as a domestic tabby on a back fence, only several times more vocal. They howl, snarl, spit, and scream like a demented woman. One night my wife and I were on the edge of a Vermont lake, gazing at the glow of a little fire on the opposite shore, when a bobcat let go in the woods behind us. No sooner had the yowls stopped echoing than we could see figures frantically running around the fire. In a few minutes the "protective" flames were blazing high enough to light up half the lake.

In late winter almost any reasonably wild area may be blessed by a bobcat. This is when the two sexes start seeking companionship. The male, normally content with a bachelor's lot, may expand his normal 10-square-mile cruising range to as much as one hundred fifty square miles during the mating season. His future mate starts on her own circle.

Occasionally each stops to deposit a few drops of urine at a convenient spot in the trail. This is, apparently, the bobcat's way of saying, "I was here and I'm available." When another cat comes along, it investigates the urine spot. Then it solemnly deposits a message of its own: "Me, too."

Eventually, then the two sexes get together, the results are apt to be deafening. Yowling and squalling, the bobcats scream sweet nothings at each other. When the tender moment passes, however, they lapse into a stealthy silence. This, of course, bodes ill for rabbits—a favorite food—as well as for mice, snakes, and ground-nesting birds.

For nearly two months they hunt together. Although they can climb well, most of their hunting is done on the ground. Often they'll try several techniques before the final one is chosen. Coming across a family of wild turkeys of roving quail, the bobcats will study the birds' line of march. Then they run silently ahead to ambush their prey at a favorable spot. Wildcats will attack deer as well—usually starvation-weakened animals during late winter—and have been known to bring down a mature buck ten times their size. The cat clings like a demon at the neck, biting for the jugular vein while the frantic prey runs and tries to dislodge its assailant against a tree or rock.

Finally the day comes for the mother wildcat's confinement. She announces the happy event to her mate in typical wildcat fashion—with a spitfire attack that sends him packing. Then, finding a suitable hollow beneath an overturned stump or under a ledge, she goes about gathering a few wisps of grass or leaves to serve as a nest.

Her mate philosophically goes back to the bachelor life he left two months earlier—"unless," as a lady biologist told me with an obvious sense of the unfairness of it all, "he happens to come across another unmated female. Then he may take up with her for a while."

The kits are born fully furred and scarcely larger than domestic kittens. A forester once showed me a wildcat den in upstate New York. It was beneath an old log on a wooded hillside. The three kits were cute, each about the size of my fist. They looked about a week old, mottled brown and buff, with bleary blue eyes just beginning to open. It was easy to see what fun they would be in a couple of weeks, hard to believe what terrors they'd be by the end of the summer.

When the kits are a few weeks old, the cat family's well-known curiosity asserts itself. They begin to waddle after their mother as she heads for the door of the den.

A friend of mine once watched the departure of a nursing female from beneath a ledge on a hillside. "It was a nice June evening," he recalled, "and I'd been sitting on a slope, listening to the birds. Suddenly I saw a movement on the opposite hill and put my binoculars on the spot. It was an old lady wildcat just leaving her den.

"Right behind her," he continued, "I could see the faces of three kittens. She stopped and stared at them and they disappeared. But when she started to walk away they showed up again. So she walked back, took a swipe at the nearest one and tumbled them all back into the hole like bowling pins. I guess they got the message that time. I watched for half an hour and never saw them again."

But they were probably back at the same stunt the next day. Wildcat babies are as irrepressible as most other kittens. As they learn to use their muscles and needle-sharp claws, they stalk each other until, with a bound, they tumble in a heap. If the mother is basking at the entrance to the den they may all sneak up on her. Each kitten's stubby little tail twitches at its own speed. Then they pounce on their "victim" with savage little growls.

At this stage they're occasionally discovered and taken home by a hiker or camper. "I've known them to make wonderful pets while they were young," a forester told me, "and rarely they may stay tame right through adult age. When I went to college in Idaho one of the professors had a bobcat that lived in his house. It had wonderful manners—even perched on the edge of the toilet bowl at the proper time.

"One man I know takes his bobcat around and lectures about it," he continued. "But bobcats have also been known to turn on their masters with no apparent warning. It's just like anything you find in the wild—it's a wild animal, and don't you forget it."

Sometime in midsummer the kits change their milk diet for one of meat. Now the mother allows the father back into her good graces for a while. Both parents may bring a rabbit or squirrel to the den—the prey only slightly injured so the kits can learn to catch it. Or they may take the family on short excursions, during which the kits stalk beetles and wind-tossed leaves with comical seriousness.

All summer the mother teaches her family the many arts of hunting. Like most members of the cat family, bobcats rely more on their eyes and ears than their noses. They learn to unthread the maze of rabbit trails in a woodland until they come to the "form" or cubbyhole where the rabbit may be crouching. A few bewildering misses show them that they cannot dash into the middle of a flock of grouse but must concentrate on a single bird. Pouncing on tufts of grass with all four feet, they carefully lift each paw to see if they have surprised a mouse. But contrary to popular belief, they seldom lie in wait in a tree for luckless animals which happen to pass beneath.

Like most cats, they have to investigate everything. Although they're largely night prowlers as adults, bobcat kittens explore the daytime world, too. They may sit in a ring and solemnly study the activity of an anthill—then just as solemnly deposit a bit of excreta on it as a fillip of their own. They may carefully turn over a pebble, roll it along a path like a domestic tabby with a ball of yarn, or dip an inquiring paw into a rain puddle. Sometimes they'll wade right into a swamp, batting away at whirligig beetles. They will take to water with less distaste than their civilized cousins. And at any moment this all may turn into frolic if they run across a few sprigs of catnip. The minty weed has the same effect on bobcats as on housecats—they roll and purr and go slightly balmy.

With the onset of winter, each wildcat youngster strikes out for itself. The parting ceremony may be just as simple case of the kittens wandering off, or it may be precipitated by a whopping family squabble. At any rate the three-quarters-grown kit, though still a year away from the adult size of thirty inches and twenty-plus pounds for males (with females a bit smaller), is nearly mature and able to take care of himself. If he lasts through the winter, he may add his own serenade to that of his parents and littermates the following spring. But though the bobcat has few natural enemies, survival is not a foregone conclusion. If he cannot find a reasonably wild area of deserted farmland or backcountry brushland to call his own, he will keep moving.

This exposes him to a number of hazards. He is quite literally caught off base. Bobcats often cover a kill and return to it. But a bobcat without his own territory must live a hand-to-mouth existence. Members of his own kind want no squatters, and tell him to move elsewhere. Thus he faces the threat that lurks for almost every creature in the wild—a shortage of elbow room.

Eventually the wandering bobcat may come to the outskirts of a farm with a chickenhouse or with a few lambs in a pasture. Such a bonanza, of course, soon brings him foul of man himself. Then he may find a price on his head. Bobcat bounties, in fact, date back to the 1700s.

To collect a bounty it was once necessary to surrender a portion of the body —often the ears—to a village clerk or other authority. But a clever trapper could cut three or four pairs of "ears" out of the belly of a bobcat. Then he'd let them ripen in the sun. When he took them to the town hall they'd be so rich that the poor clerk would hold them at arm's length—if he could get near them at all—and end up paying several bounties for one bobcat.

When word of such unlawful ways got around, it became necessary for the bounty hunter to display a bobcat nose. However, a little backwoods plastic surgery on the paws resulted in bobcats with four extra "noses"—one for each foot. Payment for bobcat tails turned out to be precisely that, with the de-tailed adult allowed to run free and raise a little of tailed youngsters.

It finally became necessary to surrender the whole animal to collect a bounty. This was fine for everybody but the poor official who found himself with a bunch of carcasses for disposal. And sometimes the easiest way to dispose of them was to take them to another official—and thereby collect a second bounty.

Such past mixups, however, were nothing compared to the state of affairs the bobcat finds itself in today. Feeling runs so high that laws may change overnight but, at this writing, the bobcat practically finds itself cursed on Tuesdays and Thursdays and blessed on Mondays, Wednesdays, and Fridays.

A few examples: several states still pay bounties on bobcats, either as a whole state or as individual towns and counties. Until recently, some Connecticut towns still offered a bobcat bounty even though they actually had nothing more menacing than a skunk at a garbage pail since the last century. On the other side, most states now recognize the wildcat as a furbearer, its hide worth up to $20 for coats, gloves, and stoles.

Nor is that all. "In most places, rabbits and wildcats just naturally go together," a game warden told me. "But sportsmen, finding evidence of an occasional bobcat attack on grouse or quail, have tended to overlook that fact. In one place in the Southwest they bountied the wildcat so intensively that they finally had to turn around and slap a closed season on it to keep a check on the jackrabbit population."

How has all the furor affected the number of bobcats? "Despite the bounty," says Jack H. Berryman, chief of the Division of Wildlife Services of the U.S. Fish and Wildlife Service, "bobcat numbers in general appear to be constant and may even be increasing over much of their range."

Exactly what those numbers are is anybody's guess. The ten thousand or so which part with their mottled coats—and their lives—in the United States each year may represent one fifth of the actual population, according to one estimate. Another guess says this is merely a tenth of the total number.

It all adds up to one fact: unless caught in a trap or treed by dogs, the wraithlike wildcat just doesn't stand around waiting to be counted.

Even its tracks are confusing. Rounded, two inches in diameter and placed in single file, they are spaced about nine inches apart and could almost be those of an outsized fox—until you suddenly realize that there aren't any claw marks to a bobcat track. Besides, nothing but a bobcat could lay down the trail I followed through a cedar swamp last winter—along a snowcovered log, up the slanting stub of a six-foot branch perhaps three inches in diameter at the tip, then out through space in a five-foot leap, the cat landing precisely on the tip of a similar stub without once clutching to maintain its balance. Then it walked nonchalantly back down to earth and off through the bushes.

Unaware of all the confusion he's caused ever since he first stared down at the forerunners of Ethan Allen, Davy Crockett, and Daniel Boone, the bobcat goes his ornery, independent way. He lends his name and his reputation to athletic teams all over the country. He sends delicious chills along the spines of campers who fancy they can see his eyes shining beyond the fire.

Those same eyes, with pupils which can open enormously to see in the dark or contract to pinpoints in the sunshine, are so keen that Indians thought the bobcat could see through rocks, trees, even hillsides. The fastidious way it often covers its excreta gave rise to an Indian belief that its urine turns to precious jewels. And so the bobcat travels steeped in legend.

Actually, chances are slim that you'll ever see a bobcat. But then you could be as lucky as two friends of mine who once saw a wildcat up a telephone pole in broad daylight on a country road. Or you could equal my experience and stumble across one that glares at you with a soundless snarl from a tussock of swamp grass—only to evaporate like a mist while you swear he's still standing there.

Which is doubtless why the French trapper gave him the name he still bears in parts of Canada today: *chat mystérieux*—that uncanny cat.

THE SOCIAL LIFE OF AN UNSOCIABLE GIANT, BY WADE T. BLEDSOE, JR. *The brown bear of Alaska, the largest land-based carnivore in the world, leads a solitary life except for a few weeks each summer when salmon runs up coastal rivers bring large numbers of bears together for a common feast. Not unexpectedly, considering the asocial nature of the participants, the fishing party at the falls provides unending drama for the scientists in the audience.*

June sunlight slanted through a break in the low-hanging storm clouds and poured down over McNeil Cove as I stood alone on the beach and listened to the fading hum of Bill DeCreeft's float plane. He had deposited me at our small camp on Alaska's McNeil River after we had waited out a five-day "peninsula blow" in Homer. Weather permitting, my field assistant, Jim Taggart, would make the trip the following day.

I was here to conduct behavioral and ecological research on the area's famed brown bears (*Ursus arctos*) under the sponsorship of Utah State University and the National Science Foundation. Service and facilities were being provided by the Alaska Department of Fish and Game.

The hum of the plane receded and disappeared over Kamishak Bay somewhere beyond the volcano of Augustine Island. With no competition now, the distant roar of McNeil Falls, one and a half miles from camp, drifted in and out on the breeze. Despite the violent storms and high tides that punish them each winter, the beaches and tidal flats and sandbars presented themselves in regal symmetry. Spring had come early along McNeil River this year, and green alder bushes extended from the camp up the mountain slopes to the melting snow.

As we flew in, I noticed two bears grazing on the sedge that grows in abundance on the tidal flats stretching southward from the camp. With the warm weather ahead of schedule, the sedge had matured and brought the bears out much earlier than usual. I postponed my plan to fish the river mouth for a dinner of king salmon and decided to venture out on the flats for a closer look at these bears.

They proved to be old acquaintances from my previous visit to the area. White, a six year-old female, was being followed by Patches, an older male. This meant that White was in estrus; otherwise Patches would never have allowed such a young or low-ranking bear in such proximity. At six years, White was late in coming into her cycle. Studies conducted by the Department of Fish and Game in other areas of the Alaska Peninsula have shown that female brown bears often breed at four years of age and sometimes as early as three. White's delayed estrus might be attributable to the high population density around McNeil Falls and the unusual social stress to which the bears are subjected.

Returning to camp to unpack my supplies, I noticed the tracks of a sow and two cubs. A sow usually breeds every three years, weaning her cubs in their third spring and breeding again early that summer. The breeding is characterized by delayed implantation; that is, although mating occurs in early summer, the fertilized ovum remains dormant until the fall, when it is implanted in the uterus and development commences. The cubs are born in January or February in a premature condition, often weighing less than a pound. They are still extremely small several months later when we first see them at McNeil Falls.

Jim Taggart arrived the next day, and after setting up our camp for the summer we began our investigation. We knew it was too early for the bears to appear in any large numbers, and we spent these first day exploring the tidal flats and nearby coves in search of familiar ursine faces. When the bears finally did arrive we would be involved in intensive daily research; now we could take time to enjoy the wilderness. We smoked a summer's supply of red salmon, feasted nightly on fresh fish, and watched midnight sunsets of incredible beauty while relishing our isolation from the outside world.

Because of the early arrival of warm weather, we expected that fishing would also begin early, and we began a daily check of the falls for both salmon and bears during the last week of June. It was on June 30 that Reggie, a thirteen-year-old female, arrived at the falls to try her luck. Other bears continued to feed on the sedge flats while making periodic checks for salmon each day. By July 5 the chum salmon had begun to run and bears had begun to emerge from the alder bushes. By the second week in July, activity at the falls was in full swing.

At the outset of the fishing season the bears are markedly aggressive in their interactions. Ten months have passed since they were last drawn in large numbers to the area around McNeil Falls, and during this time they have had little contact with other bears. While more social animals band together to obtain food or for purposes of protection, brown bears, with the obvious exception of a sow and her cubs, have no need for these group benefits. They have no natural predators and can easily find food on their own.

Aggressiveness at the falls is also the determinant in successfully competing for a choice fishing spot. An aggressive bear, as opposed to a subordinate one, likewise will retain a favorable site. This is important; the fewer times a bear is interrupted, the more fish it is likely to catch. Although aggressiveness is evident throughout the summer, it wanes somewhat as the season progresses and more and more bears arrive at the falls. Such a behavioral change is necessary if the bears are to catch as many salmon as possible, for no bear would benefit from continual squabbles over fishing sites.

Throughout the summer, fishing remains very much an individual effort. Not only do the more dominant bears occupy the better spots, but different bears fish at different times of day. The older, more wary animals restrict their activities to the hours of darkness, which, although the least productive, furnish maximum concealment. The bears' fishing activities are also affected by the changing geography of the river, for during the summer the runoff from the mountain snow ends and the river level drops. As a result, by the first week of August there may be a large number of bears fishing a small area around the main current.

Methods of catching salmon vary from bear to bear. Generally, a bear stands in or near the water and waits for a fish to swim by. When the bear sights one, it pounces or runs after the fish and pins it with mouth or forepaws or both. Early in the season, virtually every fish is consumed totally, but later, as salmon become more abundant, only selected parts are eaten. Toward the end of the season some gourmets may limit themselves to salmon roe. Nothing goes to waste, however; leftovers are eaten by gulls and the less successful bears.

Our knowledge of the brown bears' activities before and after their fishing season is quite limited, although we do know that many bears graze the extensive sedge flats of Kamishak Bay before arriving at McNeil River, and that some bears may alternate between the two areas. This summer's studies underlined how much we have to learn about the animals' behavior, for the large concentration of bears we have anticipated on the sedge flats failed to materialize.

We speculated that because the warm spring had ripened the vegetation in the higher backcountry, many cautious bears had elected to remain in remote areas rather than venture onto the more exposed tidal flats. An early ripening of berries also influenced the bears' behavior. Normally they feed on salmon through July and early August, then turn to the lush berry bushes that cover exposed ridges and slopes. This year, however, we discovered ripe berries as early as July 1. As a result, we observed only forty-one bears fishing at McNeil Falls during the entire summer. This was in sharp contrast to such previous summers as that of 1963, when Robert A. Rausch of the Department of Fish and Game sighted eighty-five bears.

But even the reduced numbers represented concentrations far greater than at other Alaskan fishing streams. On several days I counted more than twenty bears fishing the falls at the same time. Considering that brown bears are normally asocial and solitary throughout the year, such an aggregation is extremely unusual. At McNeil River, the gathering of bears also is relatively stable, according to records that date from 1963. The most regular animals have returned predictably year after year.

Thus through habit and gradual recognition of each other, the brown bears of McNeil have overcome their unsociableness to engage in an extremely social situation. Even more atypically, they have adapted to the presence of humans. For many younger bears, their first encounter came during their first summer; for the older bears, conditioning to humans occurred over many summers. During this summer it was always interesting and often amusing to observe the bears' reaction to us. While many bears, including sows with cubs, gave us only a casual glance as they sauntered by no more than ten feet away, others bolted when they spotted us. In general, younger bears at the most nonchalant. It is a rare adult male that is not extremely wary of human presence.

As I mentioned, the bears' activities at the falls are largely shaped by dominance. Usually mature males rank the highest, followed by sows with cubs, single sows, and finally the younger animals. Sometimes, though, even the largest males will defer to a sow with cubs; they are fully aware of the mother's irascibility and prefer to steer clear of her. After a sow weans her cubs she drops significantly in rank, but the cubs may, as a team, rank higher than older single cubs.

Dominance at the falls is acted out in a variety of encounters. When a mature male or female approaches a less dominant animal, the latter may simply run away. But when two animals of more or less equal rank meet at a preferred fishing spot, the encounter may last for as long as ten minutes and involve much circling, bluffing, and testing. Sometimes actual fighting erupts, but before resorting to this extreme measure the bears employ numerous "signals" to communicate their intentions. These usually include body orientations, stances, different positioning of head, ears, and mouth, and other subtle postures, the specific function of which I hope to determine in my research.

Given the atmosphere of actual or potential conflict, McNeil Falls seems an unlikely place for playfulness. Nevertheless, play was very much in evidence. While it occurred mainly among juveniles of between four and eight years, I also observed females as old as six years and males as old as nine engaging in play. Like many animals, bears primarily use their mouths and forepaws when playing. A bout usually begins with mild nose-poking and soft mouthing bites followed by more intensive activities, such as standing on the hind legs and batting each other with the forepaws or locking arms and nipping each other's faces. One huge male of unknown age was forced to play by himself, as the other bears scurried at his approach. He often resorted to lying on his back and toying with his hind feet, and once I saw him pick up a large stone and worry it about on his chest as he lounged in the water.

The ultimate success of a bear hinges on its ability to reproduce. Here, too, dominance is the key, for it is usually the larger males that are seen in consort with estrous females. Thus dominant characteristics are passed from generation to generation.

Similarly, it would seem logical that a high-ranking sow would be most successful in raising young, although numerous other factors also affect the survival rate of cubs.

I had expected to see many new cubs at McNeil Falls, and I wasn't disappointed. Five sows arrived with spring litters of from two to five, and two others returned with yearlings, bringing the total to eighteen cubs for the summer. The normal confusion that one might imagine from such an aggregation was heightened by the numerous instances of cub adoption and interchange that took place.

Often, these familial mixups resulted from the location at which a sow fished —whether her site was on the east or west bank of the river. As humans were restricted to the east side, the more wary bears remained on the west. Of the four sows that regularly fished the falls, three stayed on the west side and were constantly mixing up their cubs. The fourth, Lanky, isolated herself on our side and never lost her offspring.

Cub behavior also contributed to the confusion. Lanky parked her cubs in one spot for the duration of the fishing season, and while she fished they were content to curl up and sleep, making it easy for her to keep track of them. Three sows on the opposite bank, however, fished close together with their total of ten cubs huddled behind them. Sometimes, when one sow left the area with a catch, the entire brood, anxious and hungry, would follow her. One sow, Lady Bird, once left with no fewer than twelve cubs from four litters. Another, Goldie, even nursed her hangers-on as well as her own cubs before returning to fish.

Apparently, the cubs were unable to discriminate between their mother and another sow. And the sows, while able to detect their own cubs, were quite willing to accept the waifs, nursing and caring for them for many days before the real mother attempted to reclaim them. During these reclamation encounters, the sow would make every effort to keep her own cubs while allowing the others to make a choice. The choice was confusing for the cubs; they apparently feared leaving the new mother, who in turn kept the real mother at bay. A mother's success in regaining her cubs seemed largely a matter of chance.

Interchanges can be quite complex and may result in permanent adoptions. Early in the summer, Red Collar, who had five cubs, lost one to Lady Bird. Subsequently the cub was lost by Lady Bird to unknown causes. Shortly afterward, Red Collar lost two more cubs to Goldie, who already had two of her own. Red

Collar and Goldie then proceeded to trade cubs back and forth for the remainder of the summer, making as many as twenty exchanges in one afternoon of fishing. Eventually Goldie managed to keep two of Red Collar's cubs for an extended period, nursing them as if they were her own. But by the end of the summer Red Collar had regained one of them. Next summer it will be interesting to see whether the other cub has been permanently adopted by Goldie.

The bears had been fishing for thirty days by the first week in August, when the lush crop of berries began to draw them away from the falls. Even though the salmon run had not yet ended, only one or two bears were still fishing by August 10. All others had disappeared into the backcountry as mysteriously as they had arrived.

They will return to the falls in summers to come, for the future of the McNeil River brown bears is bright. The Alaska Department of Fish and Game has initiated a program to preserve the annual fishing season, including restrictions on the number and activities of human observers. Thus, while interested people will be able to watch and photograph the brown bears in their native habitat, the bears will be allowed to fish in peace, or at least in a state of self-declared truce.

ME AND THAT GROUNDHOG, BY C. E. GILLHAM

Hidden in the peonies, a one-time bounty hunter who killed untold numbers of wolves, bears, mountain lions, and other so-called "varmints" draws a bead on a fat old woodchuck that has been devastating his crops, ruining his garden, and undermining his barn—and is unable to pull the trigger.

It was premeditated murder. A deadly ambush, designed with stealth and caution.

Exactly twenty-seven steps from my hiding place in a clump of blooming peonies was the burrow under my barn. The groundhog would eventually appear, and vengeance would be mine.

This wasn't an effete sportsman pitting his 'scope and skill against some target dim with distance. This was a teed-off landowner armed with a magnum shotgun loaded with goose shot. That groundhog and his relatives had been eating an acre of soybeans per annum. To add insult to injury, they had progressed from field to garden to barnyard, and were now in the process of undermining my old barn.

This was the third evening that I had lurked in ambush.

In my concealment of peonies I drank in the fragrance of flowers. The grass was deliciously cool and damp. There were just enough mosquitoes for music, but not so many that I had to applaud them. I've hunted in worse places than an Illinois garden, and at worse times than a June evening. Honeybees buzzed and explored the flowers around me. A cautious, bug-eyed rabbit hopped between me and the barn and sat motionless for several minutes before starting to feed on the white clover. A robin landed within three feet of him, knowing instinctively that this long-eared animal wasn't one of those unmentionable farm cats.

Suddenly, by the foundation of the barn, a shadow bulked and moved, and a huge groundhog emerged cautiously from his burrow. My grip tightened on the shotgun. This was a very large groundhog—at least fifteen pounds. I felt some proprietary pride in that. I may not be much of a farmer, but I sure can raise big groundhogs.

For ten minutes the woodchuck crouched quietly before his den. I could see him testing the breeze with his blunt nose. Reassured, he stood on his hind legs to extend his field of vision. Silhouetted there against the white asbestos shingles of the barn, he reminded me of a brown-bellied grizzly that I once saw standing in a patch of Alaskan alders—watching for enemies just as this groundhog was doing, sensing that I was out there somewhere but not being able to prove it.

Sitting erect like that, his round earnest face studying the world about him, that groundhog was a cinch. But I held my fire, partly for the sake of my asbestos shingles, and partly out of awe for this giant groundhog.

He wasn't the handsomest animal I had ever seen, but he wasn't the ugliest, either. In fact, as groundhogs go, he was a beaut. Around his muzzle was a faint circle of whitish whiskers, and his grizzled coat was sleek. He wore the expression of a fat, nearsighted, buck-toothed kid peering through a café window at a stack of pancakes. No other animal can look so fat and so hungry at the same time.

The feeding rabbit reassured him, and the peaceful robin spelled no danger.

The 'chuck hesitated no longer. He came plumping over to within a few feet of the cottontail and began ravenously feeding on the clover. There was no hurry. At forty feet on open ground I had him cold. He didn't have a prayer, no matter what he did.

I thought of the soybeans destroyed the season before. An acre of beans is worth $100. Since my land is farmed on shares, one-third of that is mine. That woodchuck and his kin had cost me $33.33. Well, maybe not quite. Considering taxes, maybe only about $20. But there was the matter of undermining the barn. My grip tightened.

Still, tractors had replaced horses, and there wasn't any hay in the barn loft, and the barn wasn't used much anymore. My trigger hand relaxed.

The groundhog sat up, although I was sure he hadn't seen or winded me. He turned his head, and I followed his gaze. A cat stalked between the 'chuck, the rabbit, and me. The cottontail took off and we never did see him again, but the cat and the groundhog hardly glanced at each other. The cat was outclassed and knew it, and the groundhog knew he knew it, and peacefully fed on.

A jet whined overhead, letting down for the St. Louis airport twenty miles away. A freight train rattled down the grade behind the barn, leaving a stench of diesel oil and attending spews of red smoldering carbon to ignite the country-side. The groundhog never missed a bite. He was indifferent to that jet and all its affairs. And I wondered if the airplane carried any purist riflemen who invested in a fine-'scoped ordnance, and if those riflemen had ever reclined in a peony bed in the evening, watching a grandfather groundhog only a dozen rifle-lengths away. And if they had, did they shoot first, or watch for a while? Did they ponder the fact that such wild creatures are becoming premium features of our landscape, and wonder which our grandchildren will regard with the most interest—future soybeans, or future groundhogs?

These peonies are softening my head, I thought. Since when isn't a ground-hog expendable? *There was a thought.* Since when, indeed?

As a federal wildlife biologist I traveled the length and breadth of wild America. In my early days I was a government hunter for Arizona wolves, coyotes, mountain lions, stock-killing bears, and other predators. I did a great deal of killing. At the same time, poisoning crews swept the range clear of my groundhog's little cousin, the prairie dog. The world was being made safe for such dull creatures as beeves, sheep, and mortgage-holders.

We thought we were right, then, and maybe we were.

But times and conditions change, and men and animals change with them. There are now fewer sheep on the Western ranges to be killed by coyotes and bob-cats. Other countries can produce sheep cheaper. There are fewer cows on the range to be robbed of grass by prairie dogs. The South has more rainfall and grass for cattle than does the dry, rocky West, and there are people today won-dering if there isn't some room left for prairie dogs, after all. Mountain lions have been classed as game animals in most states, and their bounties removed. Even in Alaska, where I conducted extensive wolf campaigns, lobos are pro-tected by law in some areas.

Some of the old days flashed through my mind. How many hundreds of times had I lain in ambush like this, but in cedars or muskeg instead of peonies, on scabrock or rotten floe ice instead of in a hoed garden, glass-watching for grizzlies or seals instead of groundhogs? I had been paid, at least partly, by people who believed that there is no place in this world for grizzly bears. But who is to judge what belongs and what does not? Who can accuse a groundhog or a grizzly of being worthless in our scheme of things? Or even the mosquito that had just landed on my neck?

I remembered the morning on the lower Mackenzie Delta of Canada when I was studying nesting waterfowl and picked up a pintail duckling. I pulled down my headnet, held the duckling on my lap, and watched him eat one hundred and thirty-six mosquitoes that had landed on my white parka. Those mosquitoes were valuable protein to the duckling. And they were valuable to the hunter who would eat the duck-converted mosquito protein later that year, although the hunter would never know it. In places where mosquitoes are heavily sprayed, what is the effect on our bug-eating birds? Where do the purple martins go when there are no flying insects? And whose loss is greater? Ours or the martins'?

How about that portly groundhog, sitting in my garden like a brown Buddha? Is he good for anything? Don't ask me or any other farmer. Ask the nearest cottontail, raccoon, skunk, or possum. The groundhog is their landlord, just as I am his landlord. This rodent is the excavator who digs the dens for these nondigging animals and makes it possible for them to live. Many is the wily cock pheasant that has escaped into a groundhog den when hunting pressure on him became too great. Even the red fox frequently eats his benefactor, then enlarges his domicile and moves in to raise a family. I am going to kill this groundhog in a few minutes now. I am going to officially evict him from my barnyard and bean patch. And I wonder: Which will be greater, my gain or the rabbit's loss? Or will I lose, too?

For years I have watched the promotion of varmint shooting and nongame hunting—and I suppose it's okay, and I know it's fun. But I've wondered if it is something that the hunter should approach with caution, because varmint hunting for sport alone is so hard for the hunter to defend ethically.

There may be times and places when man's best interests are served by killing competing wildlife, but I'm an old traveler down *that* road, and I'm no longer sure that I know what man's best interests are. Take groundhogs. I remembered the Alaskan Arctic, where I knew the hoary marmot. He was not a pest there. The Eskimos called him *Sic-sic-puk,* and his skin was light in weight and did a fair job of keeping one warm without the bulk of caribou or hair seal parkas. I have worn marmot parkas. And in the Arctic, one does not hold any animal in light esteem if it helps clothe him. But that, I thought, was in the Arctic where there are no $100-per-acre soybeans, and here is here and now is now, and you'd better blast that groundhog. But *Sic-sic-puk* had been one of the values of my youth, and youthful values fade slowly.

A groundhog, whistle-pig, woodchuck, marmot, or whatever you prefer to call him, is strictly a vegetarian, eating grasses, clover, field crops, and native plants. Probably because of his love for green things he hibernates in winter months when flora becomes drab, and this is good. Otherwise he might eat up all living things that are not mineral or animal. Because of his great appetite he is declared a pest, and do-gooders looking for some excuse to shoot something take to him with a vengeance. Bounties have been paid for the scalps of these supposedly dull creatures, yet the groundhog makes headlines yearly in predicting the weather that is in the offing.

February 2, Candlemas Day, the groundhog is said to come out of hibernation: if he sees his shadow, he will supposedly return to his den for another six weeks of winter weather. In reality, few animals, with the exception of the bald eagle and the near extinct bison, rate much more press notice than this miserable dweller beneath my barn. But he is unaware of all this powerful philosophy. He continues to graze serenely on the brink of groundhog eternity.

The robin watched him, and barn swallows dipped within a few feet of him as the shadows grew and the sun fell behind the railroad grade. The bees had given up their nectar hunt and had sensibly gone indoors, as had most of the day creatures. It wouldn't be long now until the night shift took over. Possums would soon shuffle down the fencelines, and skunks would begin their search for grubs. Blacksnakes would come out now, for they hate the bright sun. In the blooming hedges of multiflora rose the brown thrashers and mockingbirds were settling down for the night, safe behind the banks of thorns. The scent of roses from the clouds of white blossoms was almost overpowering, enhanced by the early dew. My gun barrel was showing beads of moisture. The time had come. Good-bye, grandpa groundhog.

I had asked my wife not to come near my ambush, and not to slam the back door. But I had been gone too long, and wives grow impatient sooner than their hunter husbands. The back door opened and shut distinctly. The woodchuck sat up sharply, his round, dull face turned towards my hiding place.

"Charley, where on Earth are you? Are you going to stay out there all night?" Her voice had an edge to it.

The groundhog took off. I sat up and swung the gun on him, covering him. He was running heavily, in a straight line across open ground, going straight away from me, an easy mark for almost a couple of ounces of chilled 2s. The garden flashed through my mind, and my lost soybeans, and the prospect of more generations of groundhogs tunneling under my barn. My hand tightened around the small of the stock, ready to slap the trigger, and I was suddenly struck by the tragicomic appearance of the groundhog's broad rump as he tried to make top speed and reach the safety of his den. From either end a groundhog is comical and somewhat pathetic, but especially from the rear when his fat fanny is in full retreat.

That was one thought too many. The 'chuck vanished into the deepening shadows under the barn. I lowered the unfired gun and stood up stiffly, swatting a mosquito.

"I'm right out here," I called in the direction of the house. "I've been waiting on that groundhog, but the gun wouldn't work. I'm coming in now."

INTERVIEW WITH A SHREW, BY JACK SCHAEFER

A human conversing with a lesser animal? Preposterous, you insist. Well, insist whatever you want, but don't tell the author, who regularly talks with his fellow creatures to learn more about their ways and to get their opinions of the world at large.

"—Today. . . . Poor hunting. . . . Have to hurry. . . . Ha! over there. . . . A cricket. . . . Only a little one, mostly legs. . . . Have to do better. . . . Ha! that way. . . . Hurrah! A locust larva. . . ."

A hurrah for me too. I have been moving about in the southeast corner of the field behind my house where a half-dozen fair-sized trees annually add to the leaf mold around their bases. For at least half an hour I have been moving slowly, crouched down to bring my ears closer to the ground. I have been listening to and trying to follow thin high-pitched squeaking sounds which also are moving about, occasionally to the accompaniment of a slight rustling not so much in as under the latest layer of fallen leaves. I have had an increasingly tantalizing feeling that interpersed through the squeakings are words, phrases, perhaps short sentences, just barely beyond my ability to apprehend them.

Now at last I know I am right. Perhaps my ears have opened further. Perhaps an understanding has clicked into connection in my mind. Whatever or however, I hear him down there talking to himself. Will he talk to me?

I know what he is. Three times I have caught shadowy glimpses of him as he scurries from here to there. Something like some kind of a mouse, only smaller than any mouse, with a longer snout and a longer tail. A shrew.

"—better. . . . Nice snack. . . . Any more around? . . . Ho! What's this? . . . Blocking that hole. . . . Open yesterday. . . ."

He is very close to me, close by my right foot. I am not quite sure but think I can make out the tininess of him under a fallen mulberry leaf, peering out and up at the vast looming immensity of my bulk.

I try to keep my voice low in volume yet pitch it higher than usual. "It must be my foot," I say.

"Please move it," says the Shrew.

So. He *will* talk to me. I shift the offending foot and catch a glimpse of movement as he glides into a hole in the ground so small that otherwise I would not have noticed it. I remain motionless, wondering when and where he will reappear. A flicker of movement and he is out of the hole and back under the mulberry leaf.

"Thank you," he says. "A successful expedition. Three fat dung beetles and a millipede. I can rest a bit."

"Good," I say. "Because I am curious. I know what you are but not who you are. Obviously of the long-tailed genus *Sorex* and very likely, of the five possible species here in New Mexico, the one known as *vagrans*. But by any chance are you of our unique state subspecies *neomexicanus?*"

"The *neos*," he says, "live somewhat farther south in the state. Right on *Sorex* and *vagrans*. Correct subspecies *obscurus*. Easily identified by my darkish coloring."

"Of course," I say. "But I really have not had a good look at you. At first I thought you might be not a *vagrans* but a *nanus*. The dwarf shrew."

99

"Oh, no," he says. "In general they live somewhat farther north. And I am larger than they are. I'm really quite big. Why, I weigh fully half an ounce."

"Impressive," I say. "At least I suppose in your scale of things that is impressive. Now in my scale—" I stop. He has interrupted me with a high thin squeak.

"OHHHHH! What's that? Have they—?"

I have heard it too, a reverberating booming in the distance. "Nothing much," I say. "There's some blasting going on down in town where the urban renewal people are making wonderful messes."

"Thank heaven," he says. "It had me worried. And thank you. I appreciate your telling me. But now I'm hungry again. First tremors of starvation signals. In your scale I really am small, you know. Ratio between body weight and surface area very unfavorable. Constant heat loss. Have to eat more than my own weight every twenty-four hours to keep up adequate heat and energy."

"Wow," I say. "You had a good meal just a few moments ago. Yet you are hungry again. Is it that bad?"

"Not bad," says the Shrew. "Not bad. Not good. Just is. I'm hungry most of the time. Simple case of necessity. Condition of living. Now if you will kindly remain still while I put some distance between us. Your feet are horrendous things. Either one could flatten me into instant oblivion."

He is off into his scurrying again. I remain motionless as his squeakings dwindle away in and under the leaf mold. As I listen I know what they are; his sonar system is at work. By echolocation in the dimness of his hunting grounds he can distinguish movements and shapes enough to aid his foraging. And interspersed with the squeakings he is talking to himself. I strain my hearing as the sounds fade. ". . . Ha! elm beetle larvae. . . . Not much on taste. . . . But a whole batch of them. . . ."

Another day. I am moving about in the southeast corner of the field. I move cautiously, crouched low and listening intently, but I hear nothing except the crunching rustle of my own horrendous feet. I move on, approaching a rotting old stump by the south fenceline.

"Nice of you to be so careful," says the Shrew.

I stop, motionless, looking around. "Where are you? Why no sonar signals?"

"In this nice little hollow under this stump," he says. "Resting. Well fed at the moment. Surprised a pair of brush mice in here. Courting, I think. Female skipped away fast. Male tried to make a fight of it. Too bad for him. But my gain."

"But—but—," I say. "He must have been twice your size."

"Not quite," he says. "But plenty plump. Still enough left of him for another good meal after a while."

"You confuse me," I say. "Down there on your scale you must be a prodigious fighter, a highly efficient killer. You have just killed a fellow creature nearly twice your size. You dash about killing all kinds of things, yet—"

"I don't think of it as killing," says the Shrew. "To me it's a matter of eating. As I told you, a matter of necessity. Condition of living. You do the same. You eat meat. It has to be killed first. Do you think of the killing when eating?"

"Well, no," I say. "I get your point. I let someone else do the killing and simply buy my meat at a store. You have to be your own butcher."

"Right," he says. "An honest way of doing it. Facing up to facts of life. No offense meant, but you—OHHHHH! What's that? Have they started—?"

I need a few seconds to focus on what has alarmed him. I have heard the sound often before and am used to tuning it out. He must be fairly new to the neighborhood. "That," I say, "is just some of the kids hereabouts playing with the plank bridge over the side irrigation ditch. It is hinged on one side so it can be lifted when the ditch is cleaned. The fool kids like to get it up and let it thump down."

"Thank you," he says. "I appreciate your explanation."

"Well, now," I say. "That is precisely what confuses me. Your thanking me. Here you are living under the necessity of almost constant hunting, constant killing, activity one would think might make you a rude inconsiderate murderous-minded creature. And you thank me. You say please to me. You impress me as a polite and friendly fellow."

"Why shouldn't I be?" says the Shrew. "You've been friendly to me. Moved your foot. Made no attempt to grab me. Even knew something about me. I can't eat you. You don't want to eat me. No reason for anything but friendship between us. But pardon me a moment. My stomach needs restocking. Mouse meat digests rather fast."

I wait, considering what he has said. Not long.

"There," he says. "Fueled again. Not often a chance like this for leisurely conversation. What topic now?"

"I've been thinking," I say, "how different you are from the reputation you have. You shrews are popularly thought to be fierce, unfriendly, aggressive, bloodthirsty little beasts."

"Oh, that," he says. "We shrews long ago accepted that as another condition of living. Especially in regard to you humans when you finally came along. Except, of course, for rare ones like yourself. Your Aristotle had much to do with it. A complete simpleton in some ways. Believed without checking popular notions that we kill for enjoyment of it, of the killing not the eating, and that our sonar signals are screams of joy in anticipation of killings. But in general our bad reputation has been a good thing."

"How so?"

"I don't like to boast," says the Shrew. "But we are rather cute cuddly furry little creatures. Personally, I think more so than any mouse or hamster or gerbil. Without our reputation people would be wanting to make pets of us. Catch us. Put us in cages. Incredibly dangerous business."

"Again, how so?"

"Our food problem. We can starve to death in just a few hours. One feeding forgotten and we're done for. Maybe even before being put in a cage. Caught by someone and popped in a box. Dead by the time he gets where he's going. Oh, no. We like our reputation. The worse, the better."

"Even though it is wrong?"

"Certainly. And I just thought of something. You humans can be contra-

101

dictory. Two of your words. Shrew-ish. Shrew-d. One derogatory. Other complimentary. Impossible really to understand you humans. Much too various. Only way to take you is one at a time, individually."

"Interesting," I say. "That is scarcely the way we take all you other creatures. We tend to lump you into categories."

"Of course," he says. "That's because you are lamentably ignorant about most of us. But we do some lumping ourselves. Based on knowledge, not ignorance. For example, we lump your scientists and your militarists and your politicians into one category labeled VERY DANGEROUS."

"Now, that," I say, "requires considerable—" I stop. He is not listening. He is speaking.

"Another time," he is saying. "Important business in paw. Your poor ears may be missing it, but there's another shrew hunting not far from here. In my territory. Could be a female with romantic notions. Early in the season for that but always a pleasant possibility. Or it could be a male. If so, I'll have to get rid of him."

"Kill him?" I say. "Or at least try to?"

"Good heavens, no!" he says. "What a disgusting thought. We shrews are not mutual murderers. We'll just square off and touch whiskers, assessing each other. Then we'll try to out-squeak each other. Squeaking offers an excellent test of innate vitality and determination. I have no doubt of the outcome. I am currently well fueled with food. And this is MY territory. In these matters right adds to might."

I catch a glimpse of him emerging from under the old stump. Then I am completely at a loss. He is so quick and erratic in movement, glides through and under ground litter so easily, that I have no notion where he has gone. I listen for squeaking sounds. Nothing. He and that other shrew are out of my low limited earshot or their exchanges, romantic or bellicose, are being conducted in squeaks up out of my low limited ear range.

Another day. I have pondered his statements of yesterday and come up with questions that tease my mind. Now I am in the southeast corner of the field equipped with my questions and a small saucer containing about half an ounce of hamburger. I move slowly about, careful where I place my feet. He is somewhere near, squeaking and talking to himself.

". . . Lovely day. . . . Insects stirring. . . . Those newcomers, earthworms, wriggling. . . . All fine fare. . . ."

I crouch low, reaching out to place the saucer on the ground.

"Good afternoon," says the Shrew. "What are you doing?"

There he is, in plain view at last, not more than three feet away. He *is* a cute, cuddly, furry little fellow, momentarily almost motionless, sitting up on his tiny haunches, combing his whiskers with his forepaws.

"I am offering you food," I say. "Hamburger. Cow meat. Not your usual diet, of course, but—"

"Meat's meat," he says. "Always welcome. We shrews can digest anything in the meat line. But please do not do this often. Could corrupt me. Encourage me to be lazy. Dependent on handouts. Very unshrewlike to be lazy. Start to take life easy. Possibly fatal."

"It is not offered as a bribe," I say. "And I have no intention of trying to weaken your character. Just a friendly present of some food to fuel you for some leisurely conversation. I need information. I am still confused about you. By something you do and something you said."

He is already well into his hamburger meal, surprisingly dainty about it, taking quick small bites and chewing thoroughly. "Oh, yes," he says, voice somewhat muffled by meaty chewing. "No doubt I said too much. Sometimes do when my stomach's full. I suppose—" He stops. He sits up, small snouted head raised. "OHHHHH! What's that? Don't tell me they've—"

I have heard it too, the sudden surging sound and reverberating resonance in the air which has rattled the windows of my chickenhouse several hundred feet away. "Nothing much," I say. "Just a sonic boom. They aren't supposed to do it over this metropolitan area. But some of the pilots at the air base here think it fun to bust through the sound barrier."

"Thank . . . you . . . for . . . explaining . . . it," he says "That really shook me. So real. So possible. I've actually lost my appetite. Well, had it slacken some. I'll just rest a bit."

"Good," I say. "On the resting I mean. What just happened is one of the things that puzzle me. Loud noises startle you."

"Booming noises," he says.

"All right," I say. "Booming noises. They seem to frighten you. Make you afraid of something. Of what?"

"Bombs," he says. "Atomic bombs."

"Ah . . . yes," I say slowly. "I am afraid of them too. No one should be ashamed of being afraid of atomic bombs."

"Not afraid for myself," says the Shrew. He is his alert confident little self again. "We shrews know of fear only by hearsay. We are not afraid of anything, not even of that worst enemy, starvation. If we were, do you think we would have spread as we did long ago around most of the world into almost every condition and environment and climate however harsh, even into the Arctic Circle itself? I am not afraid. I am worried. We shrews carry a terrible burden of responsibility."

I stare down at him. "Good Lord," I say. "*That* certainly requires an explanation."

"Not an easy one," he says. "And it might offend you. But you've been friendly. I'll try. First I'll fortify myself with the rest of this hamburger."

I watch him eat, daintily yet swiftly. He finishes and proceeds to wash his small face with his forepaws and dry these on several blades of dry grass.

"I see the connections," I say, "between atomic bombs and your statement that you consider human scientists and militarists and politicians to be very dangerous. The scientists have made the bombs, the militarists have them in hand for possible use, and the politicians can give the signal for their use. But wherein lies your burden of responsibility?"

"Think back," says the Shrew. "Think back to the long ago time when the reptiles ruled the Earth. The dominant life-forms. Holding almost all available ecological possibilities on land and in the sea and in the air. Ranging from small ones up to the gigantic dinosaurs. In time we shrews were there too. The first of placental mammals. The first to achieve the mammalian mode.

To achieve the combination of warmbloodedness and furry covering and nurturing of young by the placenta within the womb and birth of them alive and feeding of them through infancy from special glands called mammae. The marsupials made a start along that path, but we achieved the full goal. We only could do it, being small and inconspicuous and feeding on insects, offering no open competition to the dominant reptiles. I doubt they had any knowledge in their small sluggish brains that we even existed."

"I follow you," I say. "Just the other day I read that the fossil remains of the oldest known mammals, six small skulls dating from the Cretaceous period in Mongolia, are similar to the skulls of you shrews of today."

"Not just similar," he says. "The same. They were shrews. That long ago memory remains in our line."

"Shrews or only shrewlike," I say. "No real difference. Your point is made. Now, the next step in your reasoning."

"Logical and simple fact," says the Shrew. "In the long swing of geologic time the Age of Reptiles waned. The dominant forms became extinct. Tremendous possibilities opened up. We shrews launched the Age of Mammals. Some of us, hedging against the future, remained shrews. Others of us radiated out into ever-continuing experiments in new adaptations. They founded species after species, family after family, order after order, of placental mammals. New dominant life-forms on land and in the sea and with those redoubtable inventors, the bats, even in the air. Oh, it has been a wondrous parade for us shrews to watch through the multiplying millennia. New forms developing, old forms dwindling away, but always the mammals, our progeny, the lifeforms we launched, inhabiting and inheriting the Earth. We have been prouder of some than of others, yet still some pride in them all. And new forms still possible on into the far future."

He pauses and his tiny form shivers with a long sigh. "Yes," he says. "For a long time we were proud of the line established by those of us early shrews who took to living in trees. The primate line. Your line."

"And still I follow you," I say. "And I am not offended. I am ashamed."

"Yes," says the Shrew. "We were the beginning. You humans are the current culmination. I am worried that you will be the end. Not just of yourselves but of all of us. Let your bombs loose and their radiation flooding the world and it is the end of us all. Even of us small, inconspicuous shrews. Our hedge against the future will be canceled. We will not be here to try again. To start it all over again."

I stare down at this tiny furry philosopher and I can think of nothing to say.

"Yes," says the Shrew. "I am sorry for you humans. In the onrush of what you regard as your cleverness and wisdom and technological prowess, you know not what you do."

A flicker of movement and he is gone.

Another day. A morning. A beautiful clean morning. The smog that has been hanging over this part of the Rio Grande Valley has temporarily lifted. The sunlight is clear and bright, and my mind is the same. I have good news for my tiny furry friend.

It is there in the lead story in the morning paper lying now on the kitchen table beside my empty coffee cup. We humans are beginning to show glimmers of common sense. Progress is reported toward an arms limitation agreement with the Soviet Union. Emphasis on limitation of atomic weapons.

I hurry toward the southeast corner of the field behind the house. I stop at the edge of the area where a half-dozen fair-sized trees annually add to the leaf mold around their bases and have already started what will be this year's addition in the form of burgeoning new leaves. I listen intently. Yes. Faintly I make out thin high-pitched squeaking sounds interspersed with words.

"... Fine day. ... Never felt better. ... Ha! that way. ... A fat slug. ... Hamburger's all right. ... Better to earn your own. ..."

Just where is he? Over there close by the corner? Eagerly I start forward, one step, two steps, three, four—and stop. A silence has enveloped me, a silence made more profound by the whisper of a wandering breeze through new leafy growth.

I raise my right foot and place it gently in a new position. I bend down and gently push aside crushed dead leaves where the foot has been. Held there, crouched, I stare down and there are tears in my eyes.

I am only a human. I know not what I do.

REQUIEM FOR A WILD BOAR, BY LOUIS J. HALLE
Cloven-hoofed prints on a lawn in suburban Geneva, Switzerland, are astonishing evidence of the presence of a wild boar, whose nightly visits will leave a deep impression on several lives before the medieval beast is tragically slain.

I hesitated to telephone my friend Paul Géroudet with a question so foolish. As one of Europe's leading field naturalists, editor of the Swiss ornithological review *Nos Oiseaux*, he is constantly having to weigh the claims of observers whose enthusiasm in the identification of rare species exceeds their skill; and this has sharpened his scepticism. My question was: "Do you think it at least conceivable that we might have a wild boar on our premises?"

"Our premises" is a bit of parkland sloping down to the shore of the Lake of Geneva, in its midst one of those great ugly villas built about the turn of the century. As we are only three and a half miles from the crowded center of Geneva, the surrounding area long ago became suburban. At one end of our lakeshore, however, is a bed of reeds, less than four acres, which has been set aside by the authorities as a nature reservation—La Réserve Naturelle de la Pointe-à-la-Bise. It is an oasis for waterfowl in the midst of a world of human activity.

During the week of November 12, 1972, something had been tearing up our lawn at night. I had not gone out to examine the damage until, on the sixteenth, a member of our household who had been brought up at the edge of a forest in the Haute-Savoie of France said it looked like the work of *le sanglier*. (*Le sanglier* is the European wild boar, *Sus scrofa*.) I had explained why this could not be—but when I had gone out to see for myself I had been shaken. There were even the prints of a cloven hoof. If this was not the work of a boar, it was devil's work. Either that or the great god Pan had returned out of the past.

After I had put my question to Géroudet there was a moment of silence. Then he replied that yes, he supposed a wild boar on our premises was at least conceivable. He added that there had been a report of one seen near a village some eight miles farther up the shore in rural France.

I also telephoned Robert Hainard, the nature artist and authority on mammals, author and illustrator of *Les Mammifères sauvages d'Europe*. Over the years he had spent weeks lying out all night in wild forested areas just to get moonlit glimpses of the wild boar and make sketches of it. He said the marks on our lawn, as I described them, did accord with the kind a boar leaves where it roots for food.

The night of the eighteenth was cold, cloudless, and still, the moon almost full. About nine o'clock I stationed myself on the second-floor balcony overlooking the lawn that sloped down to the lakeshore, and the lakeshore itself. I had waited ten minutes when suddenly a flock of fifteen swans on the shore, ghosts in the moonlight, all started to swim away from it. Half a minute later a black silhouette emerged from the reeds and moved rapidly on a diagonal course, across and up the lawn, until it became confused with the splotched shadows of the trees on the other side. Here it remained, a shadow darker than the rest, constantly changing shape. Occasionally it would move a few feet toward the house, then stop to recommence whatever it was doing. After some forty minutes of this it suddenly proceeded toward the house at a walking pace, coming directly under the balcony to show itself to me in the moonlight.

Through the weeks that followed, although my wife and I became thoroughly familiar with the apparition, we could never quite bring ourselves to believe in it—any more than if it had been Pan himself. Pan, however, belonged to the pagan world that had supposedly come to an end two thousand years ago, while this creature might have emerged from the pages of a medieval bestiary.

The medieval monster was more massive than any dog, with a disproportionately large head that was wedge-shaped and elongated, ending in a pig's snout. The body was at its highest behind the head, diminishing toward the hindquarters, from which a short, tufted, and upturned tail projected. This great mass was supported on relatively short, sticklike legs. The whole was clothed in thick coarse hair, black and white mixed to make gray—except that the feet and ears were black, as was a mane that rose like a crest from the top of the forehead to the middle of the back.

I have referred to this implausible creature as a monster, for so it was in the massiveness of its head and foreparts, its thickness all over, and the four short sticks on which it was supported. However, when it walked or trotted it did so with a quick and easy grace. What was extraordinary was the lightness of its trotting gait, as if the twinkling feet bore no weight at all. Over all the hours we spent observing it during the next six weeks we could never understand how such a heavy body could move so fast and daintily on such stiff little feet. In this respect it was the opposite of its degenerate descendant, the domestic pig.

The wild boar, which has been known to attain a weight of four hundred sixty pounds, is one of the great mammals, a creature of the dark forests that once covered much of Europe. Remembering tales of its savagery when hunted in the old days—how it tossed and tore up the dogs that closed in upon it, how it sometimes killed the dismounted hunter who tried to dispatch it with a spear—the question whether the one on our premises was dangerous crossed my mind. At first I simply assumed that any able-bodied person could outrun such a ponderous beast; but this was before I had seen it at full gallop.

Unless brought to bay and in desperate circumstances, the wild boar is not a danger to human beings today. I say "today" because it has developed the special wariness of man that all the great beasts of the wild have acquired since he invented missiles to kill at a distance, and especially since he took up firearms. Today even tigers and elephants avoid man, although they are unafraid of other mammals much bigger than he is. There is evidence, however, that wild boars had relatively less fear of man in the Middle Ages than they do today. Powerful as they are, they fear nothing else—neither dog, nor wolf, nor bear.

Late in the afternoon of November 21 I took Hainard on a tour to show him the signs left by our boar. There was not only the lawn that had been torn up here and there, but at the lakeshore an opening had been made through a screen of bamboos and, on the other side, a muddy, trampled path of hoof-prints led into the reeds where the boar presumably had the lair in which, being nocturnal, it passed its days. We returned to the house for tea, and were sitting before a window overlooking the lake, when suddenly the boar bounded out onto the lawn and stood before us. This was one of the two times I saw it in daylight, and the performance it now put on, as if for Hainard's benefit, was remarkable.

It trotted in a circle, it stopped, it walked a few steps, it bounded in a wide arc, it made galloping dashes, it turned about, it trotted again. One would have thought it was deliberately showing off its paces. Finally it made a dash around the house into a cleared place which is used as a badminton court in summer. We rushed to the dining-room window, now, to watch it from a distance of about twenty-five feet. Hainard had already seized the sketchbook he always carries with him, and he proceeded to make rapid sketches while the boar bounded about the badminton court—until suddenly it made off into the shrubbery.

In the days that followed, although I saw the boar only once, there were ample signs of it, including fresh hoofprints every day. I set out potatoes, which boars are said to relish, but they remained untouched. On December 8 I tried whole corn instead, spreading it near the corner of the house away from the lake, on the side with the badminton court. Shortly after dark, clearly visible in the light from the windows and from a lamp over our front door, the boar came to feed on it. After that, until its death some five weeks later, it came every evening after nightfall to spend an hour or two, with brief intervals of absence, feeding on the corn we put out for it. Generally it had left by nine o'clock, and I surmise that, after having fed, and after roaming about the grounds a bit, as it always did, it went back to the reeds.

I should explain that the property is enclosed by walls and fences. The only access to it, except from the lakeshore, is by way of a narrow drive between high walls which leads in from the public road; and there was no sign that the boar ever ventured along this drive, where it would have been trapped by any approaching car.

The evening of December 10, our friends Peter and Mary Thatcher arrived with a panoply of photographic equipment and proceeded to take the first of many flash photographs they were to make of our boar. They had brought a floodlight, which we attached to the house at the second-story level, and left there until after the boar's death; so that every evening, now, we and any guests could watch it in good light as it fed tranquilly a few feet from our windows. What other household in the whole world could assure the presence of a wild boar every evening for the observation of its dinner guests?

The Géroudets and Hainards came to dinner on the twenty-first, and Hainard proceeded to make innumerable sketches, as he did on subsequent evenings too, for the drawings and the sculpture he has since produced. My wife and I, and the children, who had returned for Christmas vacation, simply became addicted to the nightly spectacle, watching the beast for hours, as one might watch a gryphon on one's lawn, without getting over the sense of incredulity that remained with us to the end. I have often watched chamoix and ibex in the Alps, I know the roe deer of Europe and the white-tailed deer of North America, and I am not unfamiliar with such smaller mammals of the wild as foxes, stoats, and weasels. But I have never acquired anything like such intimate knowledge of the manners and movements of any other wild animal as I now have of the boar's.

The boar, for its part, adjusted to the circumstances of a human environment, including constant human activity. While it was feeding on the corn one could go out the front door and slowly approach it to some twelve feet away. It would lift its muzzle, pointing it at the approaching person, its ears pivoted forward. When the observer got too close it would turn lightly and make off, but not in panic, coming back to resume its feeding perhaps only a minute later. One could open the window a few feet away from it, with such noise as that entailed, and it would lift its head, momentarily alert, but not go away. One could go out the front door, slamming it, get into the car standing twenty feet from where it was feeding, slam the car door—and still it would remain, again alert but ready to resume its feeding. Only when the car's motor was started, or when a car came down the driveway, would it invariably make off, but generally to return again. The only boar in a wide world of human beings, it was adapting itself to its situation, and we thought that, for our part, we could assure its safety.

The boar showed itself to be more sensitive to sounds and, presumably, to odors than to visual phenomena. It did not mind the floodlight, and it took, at most, only casual notice when we turned lights on or off. The camera flashes would cause it to interrupt its feeding and sometimes walk away a step or two, but only to return immediately.

It tended to bypass any corn that was in a heap, preferring to pick it up where the grains lay separately. Pushing its flat snout along the grass or the gravel, its forelegs slightly splayed out, it would pick up the grains as it came to them, crunching them loudly between its gleaming teeth. The grains scattered over the graveled driveway were the same size and shape as the pebbles among which they lay, so that it must have had a remarkably sensitive underlip to separate them from the gravel as it did, picking them up as easily as if it breathed them in. All the while it would be switching its little tail from side to side or up and down. How well we came to know all these habits!

Apparently out of pure restlessness, it could not remain at its feeding place for more than three or four minutes, but would go off for a short saunter, as often as not simply ambling in a circle under the floodlight and returning. The first night I had seen it, rooting on our lawn in the moonlight, it had manifested this same habit of taking short walks at brief intervals. Except for the superficial damage it did by its rootings, and the paths it broke through our shrubbery, it was a most satisfactory boar, faithful in its fashion. Even before its shameful death its fame had begun to spread in the neighborhood and in Geneva itself. The number of people who wanted to see it seemed likely to be a problem.

The unpondered initial reaction of the Italian gardeners in our neighborhood, estimable and simple people, was that the boar must, of course, be shot. But this changed. I explained that hunting was forbidden throughout the region, that the presence of the boar was a matter of scientific interest, providing occasion for observation by such eminent authorities as Monsieur Hainard, and that if anything happened to it the case would be one for the police. This word, I know, spread to the entire Italian population of the surrounding area, the members of which began to appreciate the fact that our boar was, by its presence, conferring a distinction on the neighborhood.

Two days before Christmas we went off to the mountains for three weeks of vacation, leaving our house empty. But one of the local gardeners came every evening to put out the corn, and having the key to our house he let Hainard in on a couple of evenings to complete his sketching. On one of these occasions Hainard brought with him a photographer who rigged up powerful lights and made a movie of *le sanglier de la Pointe-à-la-Bise*. The boar's habits were now regular. It had settled into our community. It was established.

Before I come to the fatal last act of this history I should provide some background to explain the presence of such a great beast of the wilderness in the outskirts of Geneva.

The pursuit of the wild boar by hunters on horseback and armed with long spears, assisted by packs of dogs, was intensively practiced by the nobility in the Middle Ages and the Renaissance. Consequently, the species had disappeared from the British Isles by sometime in the seventeenth century. Before the middle of the nineteenth century it had virtually disappeared from Switzerland, a few probably remaining only along the German border. The species is, however, hardy and adaptable, it has no predators save man, and it is capable of restoring its numbers rapidly. The litter of a mature sow may number eight to twelve young, and she may produce as many as three litters in two years. The young are often the victims of endemic diseases, but the possibility of population explosions is implicit in the rate of reproduction. During both the world wars of this century there was a notable increase of wild boars in Europe, attributed to the temporary cessation of hunting in the wild forests that are their home.

In the course of the present century they have repopulated the Jura Mountains, north and west of the Lake of Geneva, and the forested foothills of the Alps to the south of it. Now in the 1970s, perhaps because of greatly increased plantings of corn, a population explosion is occurring in this area, as well as elsewhere. Suddenly wild boars are erupting into the farmlands of the Haute-Savoie and of nearby Swiss cantons. All around the Lake of Geneva they are being reported in ever increasing numbers, and at last the population pressure appears to be impelling individuals to adventure into the outskirts of villages and towns. Our boar, perhaps partly by swimming (for they willingly take to water), may have made its way down the lakeshore at night until it found, in the protected reedbed of Pointe-à-la-Bise, conditions of security for establishing its lair. It was pretty much trapped in its new home, confined to the enclosed area of the reedbed and the walled property that an American family inhabits; but in the fall and early winter there were still fallen acorns and succulent roots to be dug up. Then, when the food supply was becoming exhausted, we had undertaken to make regular provision for its feeding. A tacit arrangement, to which both sides were faithful, had been concluded.

The reader can hardly appreciate how attached my family and I became to our faithful boar unless he can imagine what it is to spend hours, evening after evening, acquiring a knowledge of a single individual such as one might have of a pet dog. When we returned from the mountains Sunday afternoon, January 14, we looked forward eagerly to seeing it again. The first thing I did was to put out its corn.

That evening, however, it did not come. Nor did it come the following evening, nor the evenings after, although we kept our watch regularly, the corn remained spread, the floodlight remained on. Someone who had been there reported that it had come for the corn as recently as the twelfth, only two days before our return, but no one had been around to see whether it had come on the thirteenth. Later I learned that, on the thirteenth, a neighbor had telephoned the local police station about six in the evening to say he had just heard gunshots at the edge of the reservation; but a policeman who had immediately come to investigate had found nothing.

Having learned this, on Sunday, the twenty-first, I went into the reservation, following the trampled path through the reeds that the boar had worn by its comings and goings. At the other side of the reservation, where it was bounded by a high mesh fence, there was a border of trees and bushes. Here the boar's highway came to an end. I looked all around, saw nothing, and was about to turn back—when suddenly I did see something. A couple of feet from where I stood the boar's snout projected from under a sort of low tent made by dead reeds that had been bent over. This was the lair our boar had made for itself, and there was the great beast itself, apparently asleep.

I knew it could not really have remained alseep through the disturbance I had been making, but it looked so much as if it were, in fact, merely asleep that I was seized by what I shall call a sense of prudence. What if it should wake up to find me standing almost on top of it? I daresay I waited a full minute, not moving, before I cautiously investigated further. The boar was quite dead and stiff, with blood at the nostrils and mouth.

I telephoned Géroudet, and the next morning he came, accompanied by the game warden of the Canton of Geneva and some helpers, to take the body away for an autopsy.

Our boar turned out to be a young female, perhaps two years old, far from full-grown, weighing one hundred seventy-six pounds. She had three holes in her left side. Two slugs from a gun had lodged in her lungs, while a third had gone through them and out the other side. It was only too easy to imagine what had happened.

As was her established habit, our boar had emerged from the reeds at the corner of our property about six o'clock on the evening of the thirteenth. Someone who knew the boar's habits, and that our house was empty, was waiting for her. It was easy to shoot her at such close range as she allowed. Pierced through the lungs, then, she had made her way back along the path, perhaps a hundred yards, to die in her lair.

The Geneva newspapers of January 29 reported the "Acte odieux dans la Réserve de la Pointe-à-la-Bise," and announced the offer by the Association Genevoise pour la Protection de la Nature and the Commission Cantonale de la Chasse of 1,400 Swiss francs to anyone giving information leading to the arrest of the culprit. On the thirty-first the *Tribune de Genève* published a pen-and-ink sketch by Hainard, drawn from life, of the wild boar that had been shot in a reservation for the protection of wildlife.

It was at dawn on January 22 that the carcass of the slain boar was removed. This carting away of the corpse, however, was not a satisfactory end. Something remained incomplete. Our visitor had, after all, been a refugee from the wilderness of primeval times, now so largely obliterated by our own macadamized world. Coming to us as an apparition from an immemorial past, what it represented was not just itself; nor was its death just one death. Therefore, in the evening of that day, at the hour when we had been used to looking for its visit, I played the recording of Mozart's Requiem—which served on this one occasion as a Requiem for a Wild Boar. When the long musical appeal drew to its end with the closing harmonies of the *Lux aeterna*, the finality of its departure had, at last, been properly solemnized and sealed.

TAMIASCIURUS—
THE HARVESTER, BY
HAL BORLAND

The incredible industriousness of wild creatures escapes our casual observations. But when the cones started to fall from the big spruce outside his study window—at the wrong time of year—this naturalist stopped to investigate. And once the cause was determined, he regularly watched. And counted. And calculated. And wondered.

It was November before I noticed that the Norway spruce beside the house was shedding cones out of season.

I should have noted it sooner, I suppose, but I was occupied with other matters. And it is a common trait to be so interested in what is happening down the road or across the pasture that we overlook the dooryard. Besides, this was an exceptionally heavy cone year for all the conifers and especially for the spruces and hemlocks in our area. Our big spruce was loaded, so full that the loss of even a third of them would have scarcely caught the eye. But there were cones on the grass, quite a few of them, and they were all from the current crop. They should have stayed on the tree till the next spring and summer.

I picked up one and examined it. It hadn't merely fallen. It had been cut off, with a quarter of an inch of stem still on it. Every cone I picked up looked the same. Then I remembered the red squirrel.

That tree, fully one hundred feet tall and four feet through at the butt, stands only ten feet from the house. From a window in my second-floor study I can reach out and touch one of its drooping branches. I work at the typewriter with my back to that window, but I can see what is happening out there from the corner of my eye.

I had seen a red squirrel scurrying up and down the big tree several times a day for a week or more, but beyond thinking that if the red squirrels try to invade the attic I will have to take steps, I paid no attention. I am suspicious of those red squirrels. For sixteen years we have lived under a kind of truculent truce with them. So the next morning I watched when the squirrel went up the tree, and a little later the cones began to come down. I didn't see the squirrel in action, but when it came down the tree a little later no more cones fell.

I waited to see what happened next, not at all hopeful. When we first came to this farm we decided to use the big chicken house for a carpenter shop and game room. But it had been unused for some years and the red squirrels had taken possession. We, of course, were the interlopers; but I insisted then, and still do, that as long as we stay here I shall be in charge.

So I carted out the trash from two of the squirrels' big midden heaps, six full bushels of spruce cone scales, twigs, shavings, and suchlike. I gave dispossession notice. But the squirrels ignored it. They gnawed electric cords, tore up sandpaper, stole steel wool, knocked cans of paint and jars of varnish off the shelves I built, and appropriated every Ping-Pong ball in sight. They also built new midden heaps in the corners. I kept destroying the middens. It took two years, but we finally won. The squirrels moved to the woodshed.

We didn't mind sharing the woodshed with them, though they built middens back of the firewood, made off with work gloves, balls of twine, garden markers, and rags, and rifled and scattered trash waiting to be hauled to the dump. We tried to ignore them. But one March day I took down from the wall the old washtub we used to boil down maple sap for a few quarts of syrup, and in it was a seething mixture of squirrels and midden trash.

I don't know how many squirrels were there, but they boiled out, up my arms, over my shoulders, between my legs, desperate to get away. They streaked out the door and didn't come back. I thought they had taken to the woods, where they belonged, but a few days later I saw that at least two of them had taken up quarters in big old apple trees in the backyard, appropriating holes the flickers had opened and enlarged.

Since then they have lived in the apple trees, properly arboreal. I don't consider our farmyard an ideal place for them, since the big spruce is the only conifer within a quarter of a mile, and *Tamiasciurus* is essentially a conifer-woods squirrel. But there are some advantages, I suppose. Just across the road are the big sugar maples where they go in March to cut twigs and open taps from which to drink sweetish new sap. Just beyond the maples are riverbank basswoods with tiny nuts for special autumn fare. Nearby are several box-elders, loaded all winter with fat-seeded keys, good emergency food. On the riverbank grow wild raspberries and strawberries, various viburnums, choke-cherries, and a variety of mushrooms and other fungi.

The living, I would think, is easy here. They seem to think so too. They stay.

There has been only one lapse in our truce. Two winters ago one of them, apparently hungry for the taste of the terminal buds, began cutting shoots off the big spruce. I watched, merely curious, until it had cut two bushels of those tips. Then I tried every nonviolent way I could think of to stop the squirrel for the sake of the tree. Nothing worked. When the cuttings amounted to more than four bushels I got down a gun and took drastic action. And that was that.

Now another red squirrel was at work in that tree, probably only harvesting a bushel or so of cones that it would stow somewhere, maybe in the woodshed. But it called for watching.

We had three excellent observation posts. One was my study window. Another was the south window of the library, which looked directly out on the tree and the whole side yard. The third was the window over the kitchen sink, with the same view as that from the library but from another angle. So Barbara and I watched from the library over morning coffee, from the kitchen while getting meals, from my study while I was supposed to be working, often from two places at the same time. We compared notes, summoned each other to see something special, and wasted several hours a day.

We soon began calling the squirrel Red, partly to distinguish him from the gray squirrels, partly for convenience. It never was a "familiar" name. The squirrel wanted no familiarity, nor did we. We would rather the wildlife around us stay wild, for its own safety and our own independence. And Red certainly had no notion of becoming a pet. If we went outdoors or even made a quick gesture at a window, he fled. He wanted none of us.

The first two days we watched he simply came and cut cones from the high branches. He came early, half an hour or so after sunrise, between seven o'clock and seven-fifteen. He worked about half an hour, cut down twenty or twenty-five cones, and left. The third day he began sorting out the cones and piling them at the foot of the tree. I thought at first that this was a way station from which he later would move them. It wasn't. It was his chosen site for his store of cones.

At first he was like a youngster in a candy store, with so much to choose from that he couldn't decide. He nosed cone after cone before he chose one, picked it up by its stem end, and hurried to the foot of the tree with it. He worked the better part of an hour that day but wasted so much time choosing that he didn't carry more than forty cones to his storage pile. He built that first pile in a kind of shallow alcove between the bulging shoulders of two big roots, and nearly all the cones were laid up carefully at right angles to the tree trunk.

The next day he spent less time choosing but lost a good deal of time chasing a gray squirrel that came from up the road. Red evidently had laid claim to the whole side yard as his private domain. The gray came down the road, into the yard; and Red dropped the cone he was carrying, rushed the gray and chased it back up the road. There was no fight, just the chase. Red chased the gray four times that day, twice the next. Then the gray stopped trespassing. It still came down the road but it passed the dooryard on the far side of the road.

The other red squirrels stayed away. Before Red's cone harvest began, I often saw two, sometimes three red squirrels chasing each other up, through, and down the big spruce in what looked to me like a game of tag. But after he took possession none came closer than the far end of the vegetable garden, almost a hundred yards from the tree.

Even the birds seemed to respect some kind of *Keep Off* sign, all except the blue jays. Half a dozen robins and a couple of flickers that evidently were wintering over here had been coming to that side yard almost every day to forage. After Red began gathering the cones they stopped coming, through we never saw Red challenge them. The jays, curious as always, had to see what was going on.

But nothing happened until the blue jays, perhaps merely curious again, began pecking at the cones. Then Red rushed them, drove them off the ground and into the tree, where they sat and watched until Red was busy again. Then they flew down and the chase and flight were repeated. Two or three times Red went up the tree and drove the jays out of there, but he never won the war. The best he could do was keep the jays at a little distance.

He was strictly a morning worker. Only once did we see him at the tree in the afternoon, and then he seemed to be merely inspecting. He tended to work in short bursts, twenty minutes or half an hour at most. Then he would dash off somewhere else and be gone for ten or fifteen minutes. Sometimes he went to the pear tree beside the vegetable garden, climbed it, and leaped from there to the nearby apple trees and frolicked in the high, limber branches.

I decided that his attention span was thirty minutes at most. And the day after I made that note in my journal he proved how wrong I was. It was a cold, raw, rainy day. I wouldn't have expected any animal to be abroad except for food. But Red appeared at seven-thirty, went to work, and kept at it, on the ground and in the tree, with only four brief breaks, until eleven-thirty, a four-hour span.

That was the day he started a second storage pile in another alcove among the tree's roots. Before the week was out he started a third pile. All were like the first one, close to the trunk, close-packed, almost as neat as a woodcutter's rick of cordwood. I went out after Red had quit work that day and counted the cones in his storage piles. They tallied 85, 65, and 61, a total of 211 cones.

Several things made us think Red was a young squirrel, probably from the previous spring's litter. He looked young, and through the field glasses we couldn't see a scar on him. And he was still learning. When we first watched, he carried every cone by its stem end, looking as though he had a huge cigar in his mouth. But midway of the second week we saw him pick up a cone that way at the far edge of the yard, start toward the tree with it, then stop, put it down, and look at it for a long moment. Then he turned the cone sideways, lifted it with his forepaws, and grasped it in his jaws crosswise. It obviously was much easier to carry a cone that way. He had learned something. Never again did we see him carry a cone by one end. Had he been an old squirrel he surely would have learned the easy way to carry a cone long before that day.

That afternoon I took my postage scales out and weighed twenty-five average cones. They weighed a total of seventy-one ounces, or just a trace under three ounces apiece. I couldn't weigh Red, but William H. Burt, in his *Field Guide to the Mammals,* gives the weight of an adult red squirrel as 7 to 8⅘ ounces. So every time Red carried a cone to his growing storage piles he was carrying one-third of his own weight. It was as though I spent an hour or so a day carrying fifty-five-pound bags of flour or potatoes in my mouth.

A few days later I found a cone that had been stripped of its scales. It looked like a miniature brown roasting ear from which a hungry eater had gnawed every kernel. In effect, that's what it was. Someone—probably Red, though neither Barbara nor I saw him at it—had stripped that cone of its scales, one by one, to get at the tiny nuts. This, of course, is the reason red squirrels gather cones. They find a seed under each scale, an edible "nut."

Wondering what a spruce nut tastes like, I chose a partly dry cone and tried to strip off the scales. But I am no squirrel. I finally had to break the cone in two—itself no easy trick—then peel back a few scales from the broken end. Under each scale I found a tiny black seed with a translucent vane that looked like a tiny flake of crust from the very best Danish pastry. I broke off the vane and crushed the little seed between my teeth. As nearly as I can describe it, it tasted like a hazelnut with a slight pitchy flavor. I chewed four or five seeds and got that taste from all of them. Seeds of the Norway spruce are very small and probably run nearly 10,000 to the ounce. I haven't the precise figure, but those of the Engelmann spruce run 8,500 to the ounce, those of the red spruce 8,700, those of the white spruce 15,000.

Red harvested and piled cones at the foot of that tree for five weeks. He failed to appear—at least we didn't see him—on only two days. He built five piles of cones, all in alcoves between the tree's roots. I would guess that he cut about 90 percent of the tree's current crop of cones, and he gathered and stored at least 85 percent of the cones he cut. The third week in December, when he had almost stopped work, I counted his store of cones as best I could. One of the piles had cones stacked five deep, so I had to estimate. My count, pile by pile, showed 95, 260, 102, 130, and 146, for a total of 733 cones. This works out to about 135 pounds, or at least 275 times the squirrel's own weight.

If I stocked our root cellar proportionately, I would put away something like 22½ tons.

This, of course, was open storage. But the squirrel knew what he was doing, probably by instinct. The first good snow we had drifted around the big spruce and covered those cones with almost a foot of clean, protective blanket. Every snow of the winter would add to the cover. And the store of cones would be available to Red in any kind of weather, because red squirrels are snow-tunnelers. Every winter they dig tunnels across the backyard from their nest trees to the big apple tree where we hang bird feeders. The birds spill seed onto the snow, it is sealed in, and the squirrels tunnel over, dig out the buried seed, and feast. They live well under those feeders and they are safe from hawks, owls, and foxes coming and going.

So that is what Red will do when he wants a varied fare—dig a tunnel to his store of cones. I write this in mid-January, with a foot and a half of snow on the ground, the temperature below zero, and not a squirrel in sight. Red is holed up in his apple tree, no doubt, sleeping through the deep cold. But when it moderates he will be up and out, and he will dig his tunnels. He will eat spruce seed as well as birdseed, and probably basswood seeds and small acorns. In March he will drink maple sap and race through the treetops with wild, springtime abandon. I only hope he doesn't decide he wants a diet of spruce buds. We have grown to like Red, from a little distance. We rather hope he stays around.

WINDOWS ON WILDLIFE, BY EDWIN WAY TEALE

Should you own a secluded house on an old farm that, by your choice, has been returned to nature, you need only to step to your windowsills to investigate the habits and activities of your many wild guests.

A cottontail rabbit came hopping over the drifts in the dawn of the third day after Christmas. The thermometer stood at zero. But the nightlong gale, the winds, and the driving snow were gone. We stared through our bedroom windows on a cold, white, winter world where the tree trunks were plated with snow on their northern sides and sweeping drifts curved up to the top of the walls along the lane. In all that frozen expanse the only animal life we saw was the small dark form of a cottontail rabbit.

Halfway up the lane, between the bridge and the hickory trees, it stopped, turned aside, and ascended the smooth, featureless breadth of the drift. There it paused for a moment. It looked intently about for possible enemies. Then it began digging rapidly with its forepaws. Snow streamed out behind it. In a few moments its head, its body, its cottontail, and finally its long, kicking hind legs disappeared as it tunneled in an almost vertical descent through the drift. Less than a minute went by before it backed out and carefully looked around in all directions. Then it disappeared into the hole again and the snow came flying out as before. After this had gone on for several minutes, the rabbit reappeared, cottontail first. Moving back down the drift about ten inches, it commenced digging once more. This time it vanished into the snow. Although we watched for a long time, it did not become visible again.

Later that morning I plowed through the drifts to the spot. When I peered into the two holes, I discovered that the first tunnel ended at frozen ground. But the second descended directly into the entrance of a woodchuck burrow. There, on this zero morning, the cottontail huddled snug and warm in a sheltered retreat.

Creatures that save their lives by flight, as rabbits do, are always aware of where they are in relation to places of safety. Their territory sense must be particularly keen. For when a predator appears suddenly there is no time to look around for a hiding place. This we realized. But never before had we seen a rabbit's sense of location so dramatically demonstrated as on this winter morning. Even on that white, uniform, unmarked expanse of the snowdrift, the cottontail had missed the hidden woodchuck hole by less than a foot on its first attempt and had scored a bull's-eye on its second.

Living as we do in a secluded house in the country, surrounded by woods and fields and streams, we spend the year as in a permanent observation blind. Looking out our windows we continually see things of interest. Not infrequently we witness occurrences we have never observed before. All the birds and mammals around us are existing under natural conditions. They are not tame; they are not pets. They live in a kind of symbiotic relationship with us. Our presence provides them with a certain amount of protection; their habits and activities provide us with entertainment, instruction, and pleasure. The windows of our house, facing as they do in all directions, are windows on wildlife.

I remember how, for several days one year, we were diverted by the actions of an indomitable gray squirrel. We first noticed it soon after I hung a roofed-over bird-feeder on the clothesline attached to the apple tree just northeast of the kitchen window. The squirrel had dropped down from a branch above the feeder and while birds waited in the tree it sat stuffing itself with sunflower seeds. I slid the feeder along the rope beyond the end of the branch. When we looked out again the squirrel was where it had been before. It had easily leaped from the tip of the limb. I shifted the feeder farther away from the tree. The squirrel climbed to a higher branch and bridged the gap in a longer leap. This went on day after day while I advanced the feeder farther and farther out along the clothesline. Each time I moved it the squirrel climbed higher into the apple tree. It was like a game of checkers. I made a move. It made a move.

Finally, when the feeder was a dozen feet from the tree, the squirrel landed on it with such force it could no longer cling in place. It hit the hanging feeder a glancing blow that made it bounce and tilt, and then it went spinning on through the air, twisting so it landed on its feet in the soft grass below. But it was still unvanquished. It was still ahead of the game. For each time it struck the feeder, sunflower seeds spilled out and showered down on the grass. There the gray squirrel fed until they were gone. Then it went scrambling up the tree to make another leap with the same results. Before it gave up, I had moved the feeder nearly fifteen feet out from the tree and the squirrel was climbing to the upper branches before launching itself into space toward the hanging target below. How far was it leaping? As nearly as I could calculate, the little animal, in its jumps from the limb to the feeder to the ground, was traveling at least twenty-five feet through the air.

Near this same apple tree, on a day when four or five inches of light, fluffy snow had fallen, we viewed the antics of another animal, a smaller red squirrel. It would plunge into the snow, tunnel like a high-speed mole beneath the surface, and pop up half a dozen feet away. Watching it was like watching a loon diving on a northern lake. We could rarely guess where it would emerge. The little animal, so intense in its every action, appeared to be having fun, playing a game, enjoying itself by sporting under the soft blanket of the new-fallen snow.

In contrast, there was another animal that I watched from my study window one year on the first day of March. At first glance it suggested some Jurassic monster in miniature as it toiled slowly across the ice of the pond. It was an opossum out in the winter weather. Almost at the opposite extreme from the lively red squirrel, this primitive mammal moves as though permanently locked in low gear.

Even slower and even more surprising was another and smaller creature I gazed down upon when I looked out another study window one morning toward the end of December. A black-and-brown woolly bear caterpillar crawled over the frozen crust of a drift below. I checked the thermometer. The mercury stood at twenty-four degrees—eight degrees below the freezing mark. There was no wind that day. The sun was bright. From time to time I saw the creeping caterpillar curl up and lie on its side. It would remain thus for several minutes like a woolly doughnut on the snow. Its stiff hairs lifted it above the cold frozen surface below and its dark colors absorbed warmth from the sun. It may be that the bodies of these larvae contain some "antifreeze" chemical such as has been found inside carpenter ants during the period of their winter hibernation.

Through various windows at various times we have seen such things as a crow hunting in the yard and pulling earthworms from the ground like a robin. We have seen a white-breasted nuthatch, instead of following the usual procedure and hiding sunflower seeds in crannies and crevices in the bark of trees, secrete them in grass clumps. We have seen a muskrat, in the dead of winter, struggle through the snow up the slope from the pond and feed on cracked corn put out for mourning doves. We have seen red-winged blackbirds, just as roadrunners turn over flat pieces of dried mud in the Southwest, flip over plates of frozen snow in early March to expose the birdseed attached to their undersides.

Through our windows we have noticed how woodchucks carry dry leaves into their burrows for bedding while chipmunks bring in mouthfuls of damp or wet leaves. The reason, I think, lies in the size of the holes. The burrow of the woodchuck is large enough to admit the dry, brittle leaves without breaking them while the small entrance of a chipmunk's hole will accommodate only the softened, more flexible, dampened leaves.

On the day before Christmas in 1968, a mourning dove feeding on cracked corn scattered along the lane flew up and struck my west study window in a hard glancing blow. It left behind a small fluffy feather attached to the glass by the tip of its shaft. Apparently it was cemented in place by a bit of torn-out skin or flesh that dried and hardened and became as strong as rawhide. Day after day I watched the feather fluttering first in one direction then in the other in the changeable winds. Each morning I expected to find it gone. But it survived sleet. It survived snow. It survived gales. It survived freezing rains. It remained there while I worked on five chapters of *Springtime in Britain*. Week after week I saw it clinging in place whenever I looked through the glass. I wondered each morning how long it would be before it lost its grip. When at last I found it dislodged and frozen in ice at the bottom of the pane, it was the first day of March. Soft, wet snow, packing around it, had given the gusts greater purchase, and they had torn it free from the glass. But it had remained there all but three weeks and three days of the entire span of winter.

Toward the end of August, one summer, I was working at my desk when Nellie called from the kitchen:

"Here's something I've never seen before."

I joined her at a window. On the ground beneath a plastic jug we were using as a feeder for chickadees and nuthatches and purple finches, a chipmunk and a white-breasted nuthatch were hopping about, seeking fallen sunflower seeds. Each time the chipmunk neared the nuthatch, the bird spread both wings to their fullest extent and fanned its tail wide. Then it began slowly rocking from side to side. It touched the ground first with one wing-tip, then with the other. Five or six times it repeated this sidewise swaying in slow motion. Apparently it was a threat or warning or intimidation performance. However, the chipmunk paid little attention to it. Whenever, in its hurried zigzag progress, it headed directly toward the bird, the nuthatch darted in a quick retreat into the air. Its actions were new to us, a fresh discovery. But since that day we have seen this novel game of bluff repeated on a number of occasions, sometimes before other birds but most often in the presence of a chipmunk.

It was a chipmunk that was involved in another episode observed through the glass of one of our windowpanes. One day in early June I was leaving my photographic darkroom in the basement when I heard a tapping and thumping at one of the small cellar windows. I assumed it was the brown thrasher I had seen fighting its reflection there a few days before. But when I peered through the pane, the creature that faced me on the other side of the glass was not a brown thrasher—it was a chipmunk. Like the bird, it was fighting a phantom rival. Continually it leaped at the windowpane, banged against the glass, clawed at its surface. When I went outside, later that morning, I saw the little animal, still in a rage, still battling its image in the mirror of the glass. I frightened it away. But as it ran off beside the foundation of the house, it came to three other basement windows. At each it paused and launched a fresh attack on the illusory rival that seemed to be keeping pace with its advance.

Where did the doughnut go? That was the great question remaining from one sequence of events that occurred just outside one of our living-room windows on a Sunday in February. The gusts of a winter gale pounded and shook the house. In a white smother the landscape was changing, the snow drifting everywhere. Seed I scattered in scoured patches was swept far downwind. Shining icicles, shaped by the gale as they were formed, ran along the eaves like a line of curved fangs. The weather bureau was announcing we were in the midst of one of the major storms of the century when our electricity went off. The furnace stopped. But blazing logs in the old fireplace kept us warm.

Outside, in the blizzard, most birds were lying low. But just below the living-room window, where gusts had swirled along the house and scoured the ground bare, we saw a blue jay. It was clinging with both feet to half a doughnut, one of several I had tossed from the back door. At every lull in the gale, it snatched mouthfuls of food. Each time a gust struck it, the bird tilted back, rocking with its feet still gripping the doughnut and lifting the far side from the ground. By using its stiff, outstretched tail as a prop, it caught itself at the last moment and kept from toppling backward. Then, as we looked down, a tremendous gust howled around the corner of the house. The scene was obliterated in a whirl of blinding snow. Then the flakes settled, the air cleared. But the blue jay and the doughnut were gone.

Putting on a greatcoat and mittens and pulling down the earflaps of my heavy winter cap, I went out to search for the doughnut. Although I hunted for more than one hundred feet downwind, no trace of it could I find. There was no mark on the snow showing where it had fallen. Had it been carried away in the grip of the blue jay's anchored feet when the gust had hurled the bird into the air?

Once, years ago, *Life* magazine published a painting of a flying blue jay carrying an acorn held in one of its feet. At the time I shook my head. Blue jays transport food with their bills, not with their feet. I should, perhaps, have remembered Talleyrand's observation: "All sweeping generalizations are wrong, including this one." For, on this violent day of storm, under abnormal conditions, I may have encountered an exception to the rule. The fact that a smaller piece of doughnut, of lighter weight, in an even more exposed position, had not been moved by the gust seems to rule out the possibility that the heavier piece had been carried away by the wind alone.

During several successive summers, usually in July, Nellie and I have looked out the west kitchen window and observed a mystifying rite taking place in the dust of the lane. We first noticed it early one morning when we saw a woodchuck bending down, its red tongue rhythmically appearing and disappearing. Inexplicably, it seemed to be licking the ground. For more than five minutes it remained in one place while an area about two inches square became dark with its saliva. After it had left I examined the spot. All the surface dust was gone and the harder ground beneath was scored by the animal's roughened tongue. Since then, always in June or July, we have seen not only other woodchucks but rabbits going through the same performance. Each time the animal left a little pit or depression in the lane where it had licked up the dust. On one occasion two small rabbits and an older one were all busily consuming dirt at the same time.

At first we assumed we were seeing something that, while it was new to us, was well-known to others. But when I consulted Ernest Thompson Seton's multivolumed *Lives of Game Animals*, I could find no mention of such activity. None of the other volumes in my library gave a clue to the mystery. And none of the mammalogists to whom I talked at the American Museum of Natural History in New York had ever witnessed such a performance or heard of its occurrence.

Why were these animals consuming dirt? What impelled them to lick up the dust of the lane? My first assumption was that the ground there was impregnated with the salt we had sprinkled on the stone steps to melt off the ice after winter storms. But the animals always chose the far side of the lane in preference to the ground close to the steps, the ground that would be most likely to contain salt. And we saw them licking up dirt farther along the lane where none of the salt would be present. Moreover, a friend of mine who lives in northern Maine tells me he has seen snowshoe hares eating dust along back roads in the forest, dirt roads where salt has never been applied. I wondered, for a time, if some animal smell was attracting the creatures. But when I lay down beside the place where they had left their dampened and darkened little pits, I could detect no unusual odor. All of the dirt-eaters were vegetarians. The explanation for their action that now seems most likely is that the animals find in the dust they consume some needed mineral or chemical element, some beneficial addition to their ordinary diet.

Looking out the same window through which we had watched these dirt-eaters, we were fascinated, in the early summer of 1967, by another form of wild activity. For ten minutes or more, we became absorbed in the strange behavior of a catbird.

When it caught our attention, it was balancing itself on the upper edge of the old carriage stone. This slab of gray rock, about nine feet long, five feet wide, and six inches thick, tilts up at an angle just beyond the lane to the west of the house. Originally it formed the entrance to a carriage shed that has long since disappeared. Fragments of the foundation lie among the massed goldenrod of the weed jungle beyond the tilted stone. For our near-at-home wildlife, this bare rectangle of rock forms a center of activity at all seasons of the year. In winter it provides a natural feeding tray for birds. In summer rabbits and woodchucks sun themselves on it and chipmunks appear and disappear among the tunnels they have excavated around it. It is a kind of Stone of Peace, a common meeting place for many creatures, a place of amnesty in the presence of plenty. I remember one day after I had scattered cracked corn and birdseed on the stone I looked out and saw two chipmunks, six birds, a cottontail rabbit, and a woodchuck all feeding together.

But to get back to the catbird. Its wings drooped down and its tail was spread. It staggered along the edge of the stone above a drop of eighteen inches or so to the edge of the weed tangle. It appeared to be losing its balance continually, falling to one side and then to the other, arresting itself each time with an outthrust wing. It teetered on the edge, tumbling forward, catching itself at the last instant. Minute followed minute while it continued its erratic course back and forth along the edge of the stone, looking downward all the time. The performance riveted our attention as long as it lasted.

At first sight, I wondered if the bird was anting. Then I thought it was putting on a broken-wing display. Its nest, containing young, was located less than thirty feet away in a lilac bush. Finally it occurred to me it might be in the presence of a snake. As soon as I emerged from the house, the catbird darted into the goldenrod. I moved silently up the carriage stone and peered over the edge. There on the ground below lay a coiled black snake. The parent bird, by its jerking wings and staggering movements, had been holding the attention of the serpent, distracting it from finding the nearby nest.

Since ancient times, stories of serpents charming birds have been common. I wonder if many of these tales were not derived from observing such a performance as I had witnessed. The activity of the bird catches the eye. It appears abnormal. The performer remains close to the snake instead of flying away. It seems held there, charmed, unable to leave. In reality, the reverse may be true—it is not the serpent charming the bird, it is the bird charming the serpent. By fascinating it, keeping it where it is, the bird immobilizes the snake's search for the nestlings.

A friend of mine, Edson Stocking, once told me of seeing a similar distraction display put on by a robin in a peach tree. The bird's nest was on an upper branch. When a black snake began climbing toward it, the robin alighted on a limb between the serpent and the nest. It teetered and staggered over and over, almost but not quite losing its balance. Thus it diverted the attention of the snake and led it away from the helpless young in the nest above. During a lifetime spent in the outdoors, Edson said, this was the only instance in which he had observed such a performance. It had been our good fortune to look from one of our windows at just the right time to observe this drama unfolding on the stage of the tilted carriage stone.

Three little woodchucks, born one year within a burrow hidden among the goldenrod, used the carriage stone as their playground all through the latter days of spring. There they romped, wrestled, pushed each other, fell off the stone and climbed back on again. Then, tired of play, they all would stretch out full length in the warm sunshine. One had a black face and two had white faces with black noses. The little animals were as friendly and playful as prairie dogs. In a world where there are a thousand woodchuck-shooters to one woodchuck-watcher, those weeks gave me a rare opportunity for observing their daily life. It was during these days that, from my study window, I witnessed something that I have never read about, something I observed on only this one occasion.

On the last day of June, about three o'clock in the afternoon, I saw the mother woodchuck sitting on the carriage stone with one of the young animals sitting beside her. It was now about half-grown. Just as I was turning away, she reached down, grasped it by the fur of its rump, and lifted it off its feet. It hung head-downward, partially curled. Previously when I had seen woodchucks transporting their young they carried them, as gray squirrels do, gripped by the fur of their bellies and with the little animals curled up around their necks. But in those instances the carried groundhogs had been younger and smaller.

Supporting her heavier burden with head held high, the mother started off around the house and down the lane. At intervals she paused to rest. Each time, the young woodchuck uncurled itself and, reaching down with one of its forepaws, supported part of its weight by pressing against the ground. Then it would curl up again as the mother started on. For forty or fifty yards she continued down the lane. Then, turning aside, she pushed her way into a weed tangle beside the wall.

This tangle hid the entrance to the same empty woodchuck burrow that had been the goal of the cottontail rabbit when it dug through the snowdrift on that cold December morning. A little later the mother reappeared. She came back up the lane alone. Several times in succeeding days I saw the young woodchuck near the hole to which it had been carried. It had been established in a burrow of its own. Similarly, during those days, I noticed one of the other members of the litter occupying a hole in the wild plum tangle at the eastern edge of the yard. Perhaps the original burrow had become too crowded as the young grew larger. At any rate the mother woodchuck had scattered the litter to various holes. Home ties were broken. The young animals were on their own. They all, apparently, remained where they had been taken. I never saw them on the carriage stone again.

In one of Gilbert White's letters, written from Selborne to naturalist Thomas Pennant in 1771, there is a sentence that has been described as "striking the keynote of the modern school of natural history." "Faunists," White wrote, "are too apt to acquiesce in bare descriptions and a few synonyms; the reason is plain—because all that may be done at home in a man's study, but the investigation of the life and conversation of animals is a concern of much more trouble and difficulty, and it may not be attained but by the active and inquisitive, and by those that reside much in the country."

Residing much in the country, we watch, at Trail Wood, the comings and goings of the creatures around us. We do not have to make field trips to encounter wildlife. We have only to look through our windows to see it. Through them, north and south, east and west, day and night, the year around, we can observe what the Selborne naturalist, in his quaint eighteenth-century phrase, termed "the life and conversation of animals."

Audubon's Photographers

Portfolio Two

Wading through turbulent water, swimming in deeper pools, Alaskan brown bear cubs rush after their mother as she fishes the McNeil River. (Photograph by Wade T. Bledsoe, Jr.)

In a grooming ritual that is part of the annual rut,
a great buffalo bull kneels and prods his horns into the dirt
of a buffalo wallow, rubbing his head back and forth.
(Photograph by Leonard Lee Rue III)

*The fading sun of early evening leaves the forest edge
in shadows as a bull elk bugles in a mountain meadow.
(Photograph by Leonard Lee Rue III)*

*A three-month-old Rocky Mountain bighorn lamb pauses
uncertainly as it follows its mother through a difficult
pass. Traction pads on its hooves grip the smooth rock.
(Photograph by James K. Morgan)*

Standing belly deep in an Alaskan pond, water streaming from her head, a cow moose finishes a helping of aquatic plants. (Photograph by P. B. Kaplan)

A joey nurses at the pouch of a red-necked wallaby. Wallabies are small-sized kangaroos that have colonized nearly every part of Australia. (Photograph by Stanley and Kay Breeden)

Always alert to the presence of danger, a pair of black-tailed prairie dogs pose near their burrow. (Photograph by Les Line)

WATCH
AT A DESERT SEEP, BY
SYLVIA FOSTER
A mere puddle of water in a desert canyon—a rare oasis in a wilderness that is thirsting for the rains—attracts a perpetually surprising parade of life. And attracts as well, by dawn, day, and night, a quiet and inquisitive observer to record the visitors.

On a granite boulder wedged into the canyon, at four o'clock on a May morning, I sit and wait for the dawn. Arching above me in graceful thorniness are the lacy-leaved branches of a honey mesquite. In back of me, and to my right, is a flank of canyon ragweed, whose nagging burrs imbed themselves insidiously into anything they touch. To my left, a four-foot drop-off is bulwarked by a rocky ledge and the burly, ragged mesquite trunk.

This is my hideaway, my "blind," in a south-facing desert canyon twenty miles from downtown Tucson, Arizona.

The focus of my attention, approximately thirty-five feet away in a dry but sometimes summer-flowing wash, is a seep: a permanent waterhole rein-forced by a U.S. Forest Service retaining wall. It is fringed by shrubby willows on one side and on the other by a gravelly hiking trail. Above the seep, a rock-studded bank displaying a montage of cholla, mesquite, and hackberry ascends to the talus slopes of the canyon. The slopes, regimented with saguaro, and with little-leaf palo-verde and ocotillo, end abruptly below the shorn face of a towering cliff, beneath which white-throated swifts soon will be in twittery flight.

Roughly three feet across, this puddle of water inspired me, from May to early August, to sit on the boulder and observe whatever wildlife I might be privileged to see, knowing full well I mightn't see a thing. I sat there during various daylight hours when the temperature soared to one hundred degrees or more. And at dawn and early night, face smeared with charcoal and dark clothing rubbed with creosote leaves.

Coupled with my desire to observe was the realization that this seep was one oasis, at least, in the thirsting Sonoran Desert, 1974, wherein many other natural watering sites for the desert animals had dried up. Time was in my favor. The monsoons, spilling their torrential bounties earthward, furnishing water in abundance, normally do not begin until July.

So I wait in anticipation for the dawn.

Beyond the eastern ridges of the canyon, sunlight filters across the sky, striking the cliff above me with a pink untempered radiancy. It will be several hours before it penetrates the deep recesses of the canyon, but with the light the predawn gloom recedes and the blackly silhouetted saguaros impaling the sky begin to take shape. Then the slopes. The shrubs. The willows. The seep. A pudgy, oval-shaped object with wings shuss-es by: an elf owl heading for its home in a saguaro. A great horned owl *who*s its finale as the first bird of morning, a curve-billed thrasher, whistles piercingly. What could stay asleep with such an early alarm? Now, small dark myotis bats flitter upcanyon toward their daytime roosting sites in caves and underhangs.

A series of slurred *cheer-cheer-cheer*s and a flash of vivid red rivet my attention on the seep. A male cardinal is swinging on a spindly willow branch. He drops, drinks, and flies off. He was the first: a parade of birds followed.

That morning, including the night birds, I counted thirty species at the seep and in its vicinity. Altogether in my scattered weeks of observation: fifty-two species.

Conspicuous among these were the white-winged doves, summer visitants whose yearly timing to the Sonoran Desert coincides with the flowering and fruiting of the saguaros. Depending on location and temperature, this can be anywhere between April and June. In this canyon the saguaros are just beginning to unfold their clusters of waxy, creamy-white flowers. Several whitewings fly in, their softly modulated calls preceding their audible wing-claps as they settle in a mesquite opposite me. Gambel's quail are chuckling their approach as the doves, wing bars flashing, drop to the ground, strut to the seep, and dip their silvered heads.

A Wilson's warbler, yellow-bright in the shadows, flits to the seep and plunks in for a morning bath, sloshing the water with a flurry of wings. Meanwhile an ash-throated flycatcher, mandibles clicking, is snatching insects on the wing. The chuckling of the quail, distinctly close, is followed by a r-rushing of wings as the males swoop down from their rocky lookouts.

The whitewings scatter.

Cautiously leaving the protection of the brush, the quail, in pairs, mince to the seep chuckling incessantly, males foremost, ginger heads topped with nodding black plumes. Later I would see immature young. And once only—kept so well secreted are they—five downy chicks who persisted in jumping onto their parents' backs. And in late August, after they'd reconvened, a noisy covey.

Sitting attentively, howbeit numb, I watch the quail and numerous songbirds fly in to drink or bathe. But soon a much needed cup of coffee steams fragrantly to mind. I start to rise—and freeze. Sleek and tawny-gray, a coyote is nosing uptrail. It stops once, suspiciously sniffing the air, before weaving in behind the willows. Lowering its head it laps thirstily, occasionally glancing toward the open area of the trail. Thirst quenched, it returns through the willows, climbs the bank, and vanishes amid the rocks and brush.

I clamber from my hideaway, step toward the seep, and again freeze. Dead-still on the ground, a long, striped, whip-tailed lizard has fancied a tarantula hawk for a meal, and lunging forward grabs it. The wasp's black body recoils; its wings jerk and twirr in orange agitation. But the lizard either does not have a good grip on it, or it is too large a bite. The wasp wiggles free and, apparently unharmed, angrily buzzes off. Undaunted, the lizard goes successfully for a bee. Scuttling up the bank, it beats the bee up and down against the ground and swallows it.

The seep is filled with scum, debris—an old shoe, a green comb—and algae. It is alive with mosquito larvae, giant water bugs, and other creatures with light markings on their bodies that look fluorescent as they dive to the bottom. Dead moths and butterflies float on its surface and litter the edges. Thin trails of black ants are streaming off the rocks to scavenge what they can. Bees delicately finger the mud as do wasps, numerous flies, midges, beetles, and other insects. The air sings with their buzzing and droning. A conventicle of small, blue, gossamer-winged butterflies is in progress. Gaily through the willows and over the water they flutter. On the seep's edge they congregate, wings folded, like miniature beached sailboats.

On my first excursion I had noted a sizable, many-branched prickly-pear cactus just off-trail, centered within a pile of rocks interspersed with jojoba, ragweed, and limberbush. It was considerably chewed, newly chewed I might add, for the succulent pads still were oozy. Spines cluttered the rocks and were strewn on the ground. I tried to think of what animal, among those who relish prickly-pear, might have marred this particular specimen. I ruled out such rodents as the wood rat and Harris' ground squirrel, whose small incisor teeth would give them away immediately. And jackrabbits for the same reason. Besides, while I had seen cottontails in the vicinity, I hadn't seen a jackrabbit. Which didn't mean a thing. For I hadn't seen peccaries, either. Nor their scats. Nor tracks.

5 P.M. I recheck the prickly-pear. It has not been feasted upon since my first visit; the pads have shriveled and dried.

Several yards below the seep, a ledge with an unobstructed vista across the lower desert invites me to sit and enjoy the evening. Against a gold and burnt-orange sky, the westerly mountains stand out like purple cardboard cutouts, fading soon to mauve-gray, then black. Birds are noisily proclaiming day's end and settling in their lodgings; cactus wrens chuckle, towhees scold, thrashers whistle. A hummingbird, colorless as a huge dark beetle, zooms past. Two bats flitter down the broad cleft of the canyon, trailed by others, singly or in ragged groups.

A remarkable stillness prevails, a breathing space between day and night. Then a poor-will calls, two repeated, plaintive notes far away and remote. Another calls, much closer. Now, except for a chirruping cricket, it is again quiet. Moments later an elf owl yips, followed by several more as they leave their roosting sites. Again the close-at-hand calls of a poor-will. And, unexpectedly, it is flying directly up the wash. Executing a 180-degree turn, it glides onto the sand beneath the ledge, a smooth soft weightless settling, like thistledown. As with other members of its nocturnal family, poor-wills have weak legs and tiny feet ill-suited to perching. During the day they rest lengthwise on a bough or on the ground. Expertly camouflaged in delicate browns, buffs, and grays, they are seldom seen and will flush only if accidentally disturbed. This poor-will spreads its tail like a fan, the white patches bordering each side standing out conspicuously. Again it calls, barely opening its capacious, bristle-fringed, insect-trapping mouth. Heard this way the notes are harsher, with three distinct syllables rather than two. It sits for several moments, calling repeatedly, before ascending—a mottled moth-bird against the evening sky.

Irreversibly, night pours darkness into the canyon. The clear desert sky is needled with infinite galaxies: pinpoints of light sufficient to follow the trail to the seep. I carry a four-celled flashlight, but it is seldom used, for my eyes are accustomed to darkness. Its usual purpose is to "spot" a noise.

Hiking boots on gravel are not particularly quiet. As I near the seep a "crushing" sound bursts from the willows, followed by dislodged rocks clacking down the embankment. My flashlight reveals four peccaries: rotund, squat, scrambling on short sturdy legs through the brush. They disappear before I'm sure I've seen them. Along the damp edges of the seep are their cloven imprints.

Peccaries, also called javelinas, are not pigs as we usually envision them. They are in a family of their own, and are not related to domesticated pigs, which developed from Old World swine introduced into this country years ago. Peccaries are indigenous to the New World and roam only about the southernmost regions of the Southwest, south into Mexico and South America. Pricklypear is a favorite food of these wild "desert pigs," but they are not in the least fastidious and find fruits, nuts, grubs, green plants, and even snakes equally palatable. Keen-scented but nearsighted, the salt- and pepper-colored, thickjowled, pointed-nosed peccaries look somewhat like an inept drawing of a pig. They have pink snouts that are useful for turning over loosened stones and earth, but the resemblance ends there.

Many fanciful tales have been told about the supposed ferocity of peccaries. Though unpredictable, they are not vicious. However, they will attack as will any wild animal given sufficient provocation. Upon investigation, stories of peccaries attacking people generally have proven unfounded and probably arise from the fact that these myopic animals, when alarmed, may helter-skelter in all directions at once. And if you happen to be in the way, watch out!

Nevertheless they are armed with formidable canine teeth ("tusks"), and serious injury could be incurred in a direct encounter with one. Other than man, who hunts them seasonally as game, peccaries have few predators. Most desert animals stay clear of them.

I'm disappointed that my own introduction to the peccaries is so brief. I hoist myself onto the boulder, take several peanuts from a pocket, and toss them onto a shelf below a crevice where I'd previously noted rodent droppings. If I don't see anything else, perhaps I'll see a mouse.

I don't see anything except a bat, which swoops in and skims the surface of the seep. Then a mouse finds the peanuts. Tying a red scarf over my light, I watch as it streaks back and forth the length of the crevice, between times sitting stock-still, beady eyes glowing, tiny nose aquiver. Suddenly it shoots to the peanuts, stretches its sleek body full length, and takes a good sniff. But not a nut. Back to the crevice it streaks, where again it sits motionless. Such a momentous decision for a mouse to make! Finally, deciding there must be no danger and surely tempted by the nutty smell, it snatches a peanut and bounces with it to the crevice. I can hear the shell being shucked: crunchy mouse nibbles in the dark security of its home.

3 A.M. A pale slice of moon casts shadows on the trail as I trudge upcanyon one week later. The sounds I hear are many and various: bats squeaking overhead, coyotes howling and yapping on the lower bajada. A flurry of rocks or gravel could well be a fox, peccary, or coyote; rustling leaves a large beetle or small rodent. Perhaps a snake. Sometimes I have flushed, to my regret, diurnal birds who, frightened from their perches, blunder off in shrill dismay. Noisiest of all are the night birds. The predawn hours are vibrant with yips and chitterings, hoots and *poor-wills.* One of the yips, an elf owl, swoops

down and twice encircles my head. I'm in my usual night-watch garb—dark clothing, blackened face. A red scarf with a white pattern binds my hair. Does the elf have nestlings nearby and think I'm a predator? Or do the white markings on my scarf appear as moving insects? Elf owls homed in on me on two other occasions, each time a surprise foray with the little bird uttering not a sound until, perched on a branch, it yipped and chattered with vociferousness belying its small size.

Elfs, the smallest of owls (six inches or less), roost in the tallest of cacti, the forty- to fifty-foot giant saguaros. They occupy cavities drilled by woodpeckers. There they sleep daylong until the lengthening shadows of dusk perk their appetites for a juicy spider, millipede, or insect. A mosaic of umber, sienna, and white, they are difficult to see. I've heard of stout-backed naturalists who haul ladders into the desert, brace them against saguaros, and, braving the sharp spines, ultimately locate elf owls in their holes. But it seems to me, for the average person, that ladders are an arduous if not cumbersome method, considering a saguaro may have numerous holes excavated in its trunk.

Another method is to sit patiently and let the elfs find you. A number of times I have heard them yipping against the sky, or heard their wing-swoosh, then either noted where they alighted or, because of the proximity of their calls, knew they were close by. Beneath the scrutiny of my red-scarfed light they would sit, white eyebrows giving them a glowering look. And to my amusement, but not theirs, they have even hissed, puffing and raising their wings, ready to take me on if necessary.

Late May. 6 A.M. The valley is clear to the end of the world by the time I reach the seep. A poor-will calls from the far reaches of the canyon. The last sound of night? Not quite. A great horned owl is still *who-hoo*ing as an ash-throated flycatcher bestirs and voices its first rolling *ch-peer!*

Other than a baby cottontail that hops to the seep like a furry windup toy, barely wetting its whiskers on the surface, it is another morning for the birds. Predictably the whitewings converge. It is known that these doves follow regular flight paths, habitually visiting a watering site at dawn and dusk. Conceivably, a person lost on the desert without water—an unfortunate and sometimes fatal predicament—could follow the whitewings, if able and if knowledgeable of their habits, to their drinking site and thus be saved from death by dehydration. In such a predicament I'd take this information at its worth. For each time I was at the seep, dawn or dusk, they always flew in.

The whitewings leave. Other birds pay their morning respects: a verdin flits through the interstices of the willows, snatching insects; a western flycatcher performs aerial acrobatics in the same line of duty; an Arizona vireo shyly comes to bathe; a brown-throated wren scolds my presence. Then a broad-billed hummingbird za-rooms toward me, mistaking my red scarf for a flower. Snaking through the willows, a roadrunner is busily jabbing its chisel-bill into the earth for insects and other edibles.

The gnarled mesquite opposite me, its trunk partially split by a heavy branch in a windbreak, bustles with activity. A family of four cactus wrens, two of them immature, is exploring it intimately. So are a Gila woodpecker, several house finches, an ash-throated flycatcher, and two curve-billed thrashers. They probe the shingly bark and, where convenient, work up and down the length of the crack. The young wrens, nearly full-sized, pursue their parents, begging loudly for food. If ignored, they peck at everything in sight—including each other.

A tree lizard runs up the mesquite, tailing a wren. It passes the wren, a nearly dear mistake, for the wren stabs at it. The lizard scurries full speed ahead and detours around the trunk.

June. 6:45 P.M. The seep is amorphous, lilliputian bays and lagoons of clear water inundated by mud and silt. Only bird prints etch the edge. Peccaries have not been here, nor any other mammal for that matter. As I stoop to peer at the water bugs gliding obliquely to the debris-laden bottom, I see, on the opposite bank, a slender, smooth-scaled snake about twelve inches long. A delicate ash-gray with a black head, it is a member of the genus *Tantilla*, or black-headed snakes. Secretive, seldom seen, these little ground-dwellers feed primarily on insects and millipedes. This one lacks a white collar as some species have. And while I'm sure it's a Mexican black-headed snake, I'm curious to observe it in hand. I reach across the seep, lose my balance, and as my knees go kerplunk, it reverses itself and slithers under an undislodgeable stone.

Leaving it in peace I scramble to my hideaway.

Less than an hour later a coyote trots uptrail, intent on its purpose. Although there seems to be no wind I must be "upwind," for it stops, thin nose in the air, and my scent is caught. Or is it? Processing warily it stops again, body tense and alert, tuning its large mobile ears upcanyon. It is *listening*, not smelling. Then I hear what the coyote has heard: dislodged rocks. Turning my head I see two peccaries on an opposite berm. They also have stopped, snouts sniffing the air. Reassured, the lead peccary commences surefootedly downward.

The coyote bolts and runs.

Instead of following the leader, the second peccary flops down and settles right in as though set for a twilight nap. The first peccary, full-sized and grizzled, continues past the seep. More dislodged rocks. Three more peccaries gang up on the second. It looks extremely overcrowded, wedged together as they are with the prone peccary still dozing on. A message from the gang finally nudges it afoot, and all four scramble off the berm and head seepward. Then I see two more galloping to catch up. Then a tailender. Eight altogether. I'm as thrilled as if they were elephants.

One of the latecomers is as large and grizzled as was the first peccary. Six are smaller, darker, immature. The fur is rubbed off their forelegs, as a result of their kneeling on the ground to root. These areas will develop into thick calluses as the animals mature. The six young peccaries barge through the willows, and the first in line plops into what is left of the seep; short legs jerking, it rolls from side to side. The other five crowd in somehow. Another peccary manages a backroll on the muddy edge, rudely bumping into the first, who doesn't seem to mind in the least. Meanwhile the rest are briskly scratching ears, jowls, and haunches with extended hind legs. Two stand side by side, tail to snout, snout to tail, and rub each other's rumps. All drink noisily, uttering squeaks, squeals, and other indelicate sounds like magnified stomach grumblings. Finally, mud-bespattered and content, they noisily barge back through the willows.

After the peccaries had gone the coyote returned, slipping behind the willows. But darkness had closed in, I could not see it, and I did not wish to scare it with my light. I doubt that it was able to quench *its* thirst, for the seep must have been pure mud.

I would see peccaries at the seep again indulging in similar activities, and rooting in its vicinity. Later in the season they wandered on a moonlit night within ten feet of where I was camping. One was a sow with a piglet. A diminutive, rusty-red replica of its parent, it hugged her flanks like a shadow, moving in harmony with her as she edged across the trail and joined the rest of the band feeding in the brush.

9 A.M. Three weeks later. The temperature is already approaching the century mark. No matter. We had a good rain in Tucson yesterday and I'm anxious to see what difference, if any, it has made. But rains in the Southwest are fickle things, and the storm might have as indifferently dumped its load elsewhere as at the seep. And as far as I could tell, it had. I could find no evidence of runoff, nor was the ground beneath the surface damp.

The saguaros are nearly finished blooming. Many have formed fruits which, when ripe, will split open into thirds, revealing their sweet scarlet pulp peppered with some two thousand seeds. Seed and pulp alike are eagerly devoured by birds, rodents, and other animals. The saguaro's bounty also benefits elf owls, other desert birds, and bats whose primary diet consists of insects, which are attracted in hordes to the ripening fruit.

I trudge to the seep, but again too noisily. A white-tailed deer is drinking, a beautiful two-year buck with velvety spikes. Startled, he leaps through the willows and up the bank and turns, large ears twitching, black nose glinting-wet in the sun, to give me a once-over. Then up goes his tail in last-minute uncertainty. With deliberate steps, white tail flashing, he mounts the bank and is gone.

I had seen mule deer in the foothills, and I had half-expected to see them here. But not a Sonoran white-tailed deer—a smaller cousin of the whitetails of the East—whose range today is more restricted to the upper life zones. But zonation in the desert is as variable as the plants and animals within it. The elevation of the seep is about 3,600 feet. The canyon, from this point, climbs sharply. In a mile or two the Upper Sonoran Zone—composed primarily of juniper, piñon, and oak—is reached.

I never did see mule deer, nor again this young buck who may just have been out for an excursion and found the seep handy. But I happily caught a glimpse of three white-tailed does on another morning, and with my binoculars watched their leisurely progress as they browsed an upper slope.

2 P.M. Belatedly I realize the boulder-perching doesn't have to be uncomfortable. From my daypack I pull out, put down, and sigh with relief onto a small feather pillow. And again I watch the seep.

A chipmunk with a bushy tail edged with cinnamon, clear facial markings, white behind its ears, and indistinct stripes has conjured itself at the seep. To drink, or to play? It whizzes through the willows and rock-hops across the water, repeating this performance several times. Then, on a flat-planed stone angled into the water, it splays itself, tail spread behind, and drinks. This chipmunk is joined by a second. They touch noses, tails swaying, before chasing each other under and over exposed willow roots, up and down lowslung branches, back and forth across the rocks. Calling a momentary time-out, both pause to drink. Then merrily they bound off, one behind the other, and with utter disregard for the precipitous ledges, swing themselves over a rocky overhang and out of sight.

These charming little animals are cliff chipmunks. Habitually they are found, as are white-tailed deer, in the upper life zones. Whether they are displaced chipmunks I do not know, but it is doubtful they were aware they had crossed any boundary line. I frequently saw them scampering with agility and grace on the rocks, basking in the sun, and pulling down grass stems to garner the seeds. Once I watched a cliff chipmunk with a portion of saguaro fruit in its paws. Sitting contentedly on a sun-warmed rock, it nibbled on the pulp as one might a juicy peach. Every time the sweet fruit started to drip the chipmunk turned it over, catching the drip before it dropped. Then I accidentally snapped a twig. Off the chipmunk whirled with the pulp falling apart in its mouth, trailing bits and pieces between its legs.

July. 4 P.M. Under the promise of rain I head for the canyon. Traveling darkly come the thick gray storm clouds. As the pressure gathers the wind is sensed like a malignant presence before it is felt. Then, in powerful bursts, it flays tree and shrub, propelling leaves, loose bark, and debris across the sky. The saguaros sway with a swishing like surf on a sandy beach. The wind eases, followed by dabs of rain, like wet paint, splattering the trail and plunking on rock. I reach the seep and hunch beneath an overhang. A granite spiny lizard darts about, apparently unperturbed. But the pellets become too much even for a lizard, and it scurries for cover.

The clouds disgorge the rain in slashing torrents, jackknifed with lightning, and regurgitating thunder which splinters into a thousand echoes and seems intent on jarring the rocks loose about me. For nearly ten minutes I can't see a thing. Then, with a few more plunks, the rain stops. Just like that. Is it over? I hear in succession a black-throated sparrow, cardinal, and gnatcatcher burst into chips and chirps. Taking them at their song, I venture out.

The clouds are grumbling northward. Above the canyon the sky is clear. The ground is steaming. And it is hot.

The recent heavy rains are evidenced in the canyon's flora. Prickly-pears have fattened; many have ripening fruit. Ocotillos have emerging green leaves shooting the length of their wandlike branches. Desert ferns are pushing green fronds from brown stubble. The saguaros have plumped out, a positive indicator that moisture has been sufficient beneath the surface. But not a fruit remains. Those that have not ripened and fallen naturally have probably been battered loose by the winds and pelting rains. The cacti stand shorn of edibles. Beneath their bases split pods are littered. Many with remnants of ripe fruit; others gleaned and dried like well-cured magenta leather.

137

The seep again is a small puddle of water. Insects are plentiful and now include water striders, a few boatmen, orange dragonflies, and a colorful assortment of butterflies. I follow the wash and find a number of basins in the rocks which have caught and retained rainwater. One, at the base of a narrow chute, has newly emerged, velvety black planaria gliding sinuously over the surface and along its edges.

Numerous millipedes are humping on the damp earth as I cut up a bank en route back to the seep. I hear a series of weak *bleep*s. Locating the source, I stand utterly transfixed. A full-sized garter snake has caught a spadefoot toad by its hindquarters. The toad's rear legs already have been engulfed, and the snake's upper jaw is anchored beneath the toad's belly, its lower jaw on the toad's back.

For over an hour I watched as the toad, eyes blinking, throat pulsating, *bleeped* and struggled. But the snake relentlessly "walked" its loosely hinged jaws upward, first one side, then the other, stuffing that fat toad into its monstrous mouth. As the toad's body became compressed its upper torso swelled and seemed about to burst. Several times snake and toad rolled over and over. When the snake reached the upper portion of the toad's stomach it reclamped to the right. For the first time I saw blood oozing from its body, reddening the snake's head. It bleated continuously, flailing its forelegs.

Upward moved the snake until the toad's stomach was engulfed. Now only the forelegs, throat, and head of the toad remained. The snake chewed on the toad's throat. Then, in a quick move, it reclamped its jaws on the toad's left leg. The leg disappeared. The toad's eyes still were blinking, its throat pulsating, as the snake took another hold. Then the toad gave a final weak *bleep* as its head vanished into the snake's mouth. Its right leg was last to be swallowed. I watched the tiny foot go down as a drowning swimmer's hand underwater.

I picked up the snake. Its upper body bulged with ingested toad. It might as well have been sawdust. But one hungry garter snake had had a full meal of one unwary toad.

5 P.M. Two weeks later. There has been no rain lately, although storm clouds are massing on the horizon this afternoon. Most of the water basins have dried up, including the one where planaria formerly swarmed.

It is very quiet, for there is little activity in the oppressive heat. A female cardinal is perched on a branch, beak open, wings extended.

Braving the heat for a drink, a rock squirrel comes to the seep. Its two-toned coloration blends perfectly with the surroundings; it could well be just another smooth stone lining the seep. Loudly squawking the squirrel's presence, an Arizona jay, another renegade from the Upper Sonoran Zone, flaps to a willow branch. It continues to protest vehemently, but the squirrel ignores it. After the squirrel sidles away, the jay drops in a blue flash to the seep.

As the canyon eases into shadow, a rufous-crowned sparrow enjoys a twilight bath. Two great horned owls begin exchanging notes for the night. First the male with a resonant, three-syllabled *who-hoo-hoo*, followed by a reverberating echo, then the female answering, lower pitched, softer.

I hear a fluffing, flapping sound and peer intently at the seep. An animal shaking itself? It is an elf owl in sporadic flight after insects.

A gibbous moon pushes itself with timeless reassurance over the ridge, flooding the canyon with light. But my only visitors are those nasty little midges, no-see-ums, which manage to find any unrepelled millimeter of my skin and cause furiously itching welts that last a week.

Weary of the midges I head home, walking briskly but alert. Hence I'm not particularly surprised, or alarmed, when I hear a distinct b-uzzzz-z. To anyone who has heard it before it is unmistakable, and unforgettable. Rattlesnake! My momentum carries me a few steps forward. Turning, I train my flashlight on a coiled tiger rattlesnake. Thirty inches long, buff colored with tiger-stripe cross-bands, it is a midget compared to western diamondbacks, which may attain a length of eighty-nine inches. But it is nothing to fool with, for its venom is one of the most potent of any rattlesnake. Squatting on my haunches I watch it very cautiously. It stares with glittering, blinded eyes into the light, then languidly, gracefully uncoils and slips into the night.

But this doesn't happen often. Rattlesnakes are not encountered nearly as frequently as most people believe.

August. 2 P.M. Heavy rains this past week which induced an extravaganza of summer bloomers on the lower desert have left their imprint here. Many of the herbaceous plants and vines are rampant with luxuriant growth. Wild zinnias enliven the green, as do morning, glories, globe-mallows, and *Allionia* trailing its purplish flowers on the ground.

The lower canyon is still dry, but as I near the seep I hear water: running, flowing, gurgling water. The creek is flowing! The seep is an overflowing pool five to six feet across. Water rushes and foams in from above, out from below, swiftly traveling for at least seventy yards over the wash, down a ledge, and beyond. I rush up into the seclusion of the wash. The pools are large, clear, invitingly cold in the searing heat. Jubilantly strip and jump in thigh-deep.

In less than a week there is barely enough water to splash over my face. A few isolated pools remain, dribbling over the rocks into silted areas and encouraging lush grasses, ferns, and club moss. Willow catkins blanket the surface of the seep. I sit on my boulder for several hours, a lone swallowtail butterfly and a few birds for company. I know water is freely available in the canyon's upper reaches, and in other creek beds and watering sites, so the chance of seeing wildlife is greatly diminished. But I do not mind. I'm as grateful for the summer rains as the desert animals must be.

Below the seep, beneath a shelf in the wash, is an evaporating pool rank with the smell of decaying algae. It is seething with black tadpoles: this season's promise of spadefoot toads.

Once more I visit the canyon. The whitewings have gone but black-headed grosbeaks, warblers, and phainopeplas have moved in. Green grasshoppers high-vault from the ragweed as I brush against it. The riddled leaves are thick with molting black-and-amber tortoise beetles.

Baby lizards, no longer than matchsticks, are zipping here and there, almost faster than the eye can catch them. Basking on a stone on the trail is one little lizard different from the others. I scoop it up. It fits like a small round button in the palm of my hand. A baby horned lizard, pale as sand, it is an exact replica of its fierce-looking parents, complete with tiny horns.

The pool beneath the shelf, formerly cramped with tadpoles, is completely dry. Only with my magnifying glass can I distinguish between the desiccated bits of tads and the debris stuck to the gauzy, bleached-gray algae. Did any of their numbers survive? My eye catches what appears to be a large sand flea, then another and another, jumping to the seclusion of the rocks. They're toadlets! The rains provided the water necessary for their propagation. But as so often happens in the arid Southwest, the pool had dried before all the tads had developed. Even these few vulnerable toadlets, prey to birds and reptiles, will not mature and burrow, as spadefoots do, into the ground to pass the time awaiting next summer's rains. But this was only one pool. Among how many?

There is so much I have not and may never see, even with a lifetime of boulder-sitting. The very existence of wild animals, in a narrowing world pushing them ever farther from their natural habitats, some to the verge of extinction, continues to amaze me. To observe any wild creature, any at all, not behind bars and enclosures, not behind the scope of a rifle, not half-tamed, begging for human handouts in state and national parks, but free, unstalked, unpampered, in its own environment, is a rare and beautiful privilege.

My watch at the seep has just begun.

"STRANGE POWERFUL BIRDS, SIR," BY FRANKLIN RUSSELL

A rookery, a birder will explain, is a breeding or roosting place for colonial birds of any species, although the term obviously originated with that gregarious Old World crow, the rook. While they are seldom used, the ornithological language also contains a few very descriptive and poetic words for large gatherings of nesting birds—penguinery, heronry, gullery, ternery, puffinry. Here we visit the gannetry, on a cliff high above the ocean's riches, where thousands of these regal aerialists gather annually to restore their race.

The time is dawn and the place is Bonaventure Island in the Gulf of St. Lawrence. The cliffs of the island face east, and if you stand on top of them the rising of the sun is like yellow fire exploding out of the sea. The sea and sky go black for a moment at the sun's abrupt appearance, and the gannets are like white crosses etched on the dark backdrop. From the amphitheater of the cliffs, the sea becomes a stage and the players are gannets, thousands of them gliding, diving, swimming, hovering.

My first sight of a gannet was of a solitary bird skimming the storm waves of Cook Strait, which separates the two main islands of New Zealand, but my most memorable view of one was at Bonaventure. I was walking quickly along the track that meanders through Bonaventure's woods to the gannetry, hoping to catch a glimpse of the birds before night. The woods were totally enclosed and gave the illusion of premature night. I turned a bend in the track and saw light spilling toward me. A rectangle of blue sky appeared, livid against the woods' gloom. A great bird hung suspended in the center of the rectangle, motionless as oil on canvas, but a *living* picture. The bird seemed to be waiting for me.

I put up my binoculars and he expanded into detail. His long, solid beak, shaped like the head of a Norman pike, flowed smoothly into his head, which was white flushed with gold. His white body narrowed to a wedge-shaped tail. His wings, white at the base, became black just after the joint. The black trim was perfect art, transforming an elegant white bird into a regal flying creature.

I walked to the top of the rise. The rectangle of light widened into green slopes running down to the gannetry and a great blue sea. The gannet, twenty feet overhead, looked down at me and then turned the flat of his wings to the wind and allowed himself to sweep away downwind where he dwindled to a speck.

The flight of the gannet is so distinctive that it resembles nothing else. It is not the buoyant glide of a hawk or eagle, nor the aloof circling of a raven; it is unlike the flight of any other seabird. The glide is a rush of smooth, almost liquid, motion, as though the bird were fixed on invisible rails and impelled by a silent jet force. The flight perfectly expresses kinetic energy.

Audubon found the flight of the gannet "elegant," and the word best describes a sureness, a grace that makes gulls look clumsy by comparison. The gannets pump along for a few powerful strokes and then glide, their pointed wings slightly anhedraled. Like gulls, they frequently fly purely for the pleasure of flight, particularly during boisterous weather. Once, in a Newfoundland gannetry, I watched several thousand gannets take off in a gale and use the force of the wind, acting on the upper side of their wings, to hurl themselves down from cliffs at tremendous speeds. Then, very low to the water, they swooped up and disappeared into the overcast. Merely to watch them gave me that roller-coaster feeling in the stomach. It is hard not to infer similar feelings in the birds themselves which, at the highest speeds, were likely moving at much more than one hundred miles an hour.

On another occasion, at a Gulf of St. Lawrence gannetry, I saw a second expression of the exuberance of flight. A southerly storm had built up ten-foot waves, and gannets took off in a steady stream, planing down from the nesting cliffs until they were skimming the waves. When a suitably large wave loomed up, they angled down, smashed into the sloping side of water. In a second, they popped out from the far side. There some of them bathed, flapped wings, preened, shook their feathers. It looked like great fun.

During any storm, it is exciting to watch a gannetry. It is also a rare sight, since a storm usually means you are stranded on the island. In high winds, all the birds turn to face into the wind, their bodies streamlined like teardrops. In such a position, it takes a hurricane to tear them loose, yet they often fly in such weather. Once, in a seventy-mile-an-hour gale at a gannetry, I found myself surmounted by a canopy of thickly clustered birds, their long pointed heads piercing the wind like darts, their gargling calls whipping past like bullets. Reaching up, I could almost touch the lowest birds.

The flight of the gannet is its distinction, but its dive is its drama. The dive has a quality which surely stirs poetic images, yet I cannot recall ever having read a lyrical line about the gannet. I am sure Shakespeare never saw a gannet. He would surely have included its dive to make a literary point, as he did with the habits of gulls, ravens, hawks, finches, geese, crows, and many other birds. He lived between the sites of two modern gannetries which were likely active in his time: one at Grassholm in the Bristol Channel, the other at Bempton Cliffs on the northeast coast of England. Yet because the gannet almost never flies overland, poet and bird probably never met.

The dive of the gannet is a magnificent sight—and deceptive. It looks like a simple free fall from between twenty and one hundred feet but, in fact, it is as carefully controlled as the fall of a dive-bomber. I once saw the mechanism of the dive when I was walking along some high Newfoundland cliffs. I found myself alongside a hundred or so gannets as they made a mass attack on a school of fish. Looking down, I saw the school in the shallows, perhaps capelin prowling the coastline in search of a place to spawn. As I watched, the gannets circled lower. The British ornithologist J. H. Gurney once concluded that gannets adjust the height of their dive to suit the depth at which the fish are swimming. The fish in the bay were near the surface, and the gannets seemed to know exactly how far they had to descend before diving. From my vantage point I looked down on them, almost peering over their shoulders. They began diving.

Seen at sea level, the diving gannet appears to close its wings three-quarters and fall freely before snapping the wings cleanly against its body as it strikes the water. But from above, I could see that the fall was actually subjected to several minute adjustments of the wings and tail as each bird fell. A bird could control its strike point precisely. Because the prey was so visible, I could see how the gannets selected their targets. They hit wherever the fish appeared thickest. In one moment, when a mass of fish moved close to the surface, a score of gannets hit the target area of not more than twenty or thirty square feet. Despite my closeness, I could not see or understand how the birds avoided smashing into each other, particularly since their fishing technique is to dive *below* the prey and then swim up from beneath for the attack. As birds surfaced, swallowing fish, other birds were still diving, hitting within inches of their floating comrades. But how did they avoid the submerged gannets? It was an extraordinary demonstration of flying skill and coordination.

The gannet, because of its regal appearance in the air, has always excited the admiration of bird-lovers and the grudging respect of professional fishermen and coastal folk. An old Newfoundlander who had used gannet flesh for more than half a century to bait his codfish hooks summed up the feelings of many fishermen towards a bird they admire for its grace and hate for its success as a fish hunter: "Them gannets, they be strange powerful birds, sir."

The gannet is a myth-maker. For centuries in the Old World, the gannet shared with the storm petrel and the gull the reputation that it embodied the spirits of drowned fishermen. Towards the end of the nineteenth century, an English naturalist, P. H. Emerson, found that British coastal folk believed in the transmigration of men's souls to birds. Old fishermen turned into gulls when they died, he found, and youngsters turned into kittiwakes ("They have not to be so artful," as one old man put it). The women "don't come back no more," because they had seen trouble enough. The gannet, however, seemed to be a more specific possessor of men's souls. Individual birds often represented specific fishermen returning to watch over their families. Emerson found one fisherman who watched every gannet in sight because his brother had foretold that he would return as a gannet after his death, adding that he would be recognizable by his "black armsleeves," the black-tipped markings on a gannet's wing.

One of the most enduring myths about the gannet is that fishermen used to tow flat boards behind their boats to which fish were fastened. Gannets saw the fish and dived, breaking their necks on the board. The myth is persistent and is repeated by Scots, Irish, Icelanders, Newfoundlanders, Nova Scotians, and Québecois. It's a nice little folk story but it's thoroughly debunked by all ornithologists.

The gannet is globally successful. "I know of no other bird with so few formidable enemies," Audubon observed. Gannets occupy the eastern North American coast as far north as Labrador. There are populations in Iceland and throughout the North Sea. A subspecies lives along both coasts of southern Africa and nests in South Africa. I have seen the familiar gannet dive along parts of the southern coast of Australia (another subspecies) and at many points around the coast of New Zealand.

On Christmas Day 1769, some of Captain James Cook's men from the *Endeavour* shot many gannets in New Zealand waters, and the entire ship's company feasted "in an old-fashioned way," as Sir Joseph Banks, then president of the Royal Society, described it, on "goose-pie." Cook's men must have had tough stomachs, or Banks was no epicure, because gannet flesh is reviled for its toughness and oiliness by fishermen throughout the Northern Hemisphere.

The gannet, however, is but one member of the *Sulidae* family, which includes the tropics-roaming boobies. They closely resemble the gannet in both appearance and behavior, and they puzzled Audubon with their name because he did not find them at all "stupid." The masked booby ranges the Caribbean, the mid-Atlantic, the coast of Africa, and the central Pacific almost from Japan to Australia. The blue-footed booby is seen from the west coast of Central and South America to the Galápagos Islands. The brown booby, even more widespread than the masked, is a great wanderer, traveling up to four thousand miles across open sea. The red-footed booby is almost as widespread.

All the boobies have common habits but each has unique traits. I have seen boobies planing down at high speed to overtake and catch flying fish. The masked booby may often be seen perched nonchalantly on the back of a basking sea turtle.

All boobies and gannets are inquisitive, none more so than the red-footed booby. Fascinated by ships, it attempts to land in rigging to investigate and sometimes gets itself into a fearful mess among radio lines, ladders, and Sampson post stays. It often lands on the decks of ships, particularly tankers, and then cannot fly again because hold covers, ventilators, and ladders prevent a free-running takeoff. Once, when I was boating off Bonaventure Island with John Paget, the game warden of the island, hundreds of gannets began leaving the towering cliffs and collecting over us. I am well accustomed to being menaced, or even attacked, when in the vicinity of seabird bazaars, but this was an attention that was discomfiting for another reason. The gannets gathered over us in absolute silence, about three thousand of them, using a slight southeast wind to hover in a thick and ultimately intimidating throng over our heads. Paget gestured to them and shouted. A thousand birds dropped suddenly lower. We could see their gleaming, intent eyes. It was like being watched by a multitude of very intent and absorbed children.

One final distinction of the gannet is its various breeding environments. Each, in my experience, has a special atmosphere. Bonaventure is perhaps the grandest, certainly the most accessible, but Cape St. Mary's, a southeastern tip of land in Newfoundland, has the most spectacular colony. After a seven-mile walk across peat barrens so bleak that one ornithologist declared it would break his pen to describe them, the gannetry appears as a splash of white about the size of a tennis court, on top of a spire of naked rock. Professor Wynne Edwards, the English ornithologist, saw the thousands of hovering gannets as "a cloud of snow flakes poised over it." Only one man has ever set foot in this gannetry, John C. Cahoon, an American who spent some time in Newfoundland collecting specimens. He was so shaken by the climb that he had to be rescued by ropes thrown to him. He was later killed in a fall from a cliff on the same coast.

The gannetry on Funk Island, a tiny speck of rock about fifty miles northeast of Newfoundland's coast, is another spectacular colony but for different reasons. The gannets, lacking the protection of cliffs, have displaced the murres from their nesting places on the highest piece of ground on the island, and so appear as a livid patch of white buried in the black mass of hundreds of thousands of murres.

The other principal gannetries in the eastern Atlantic are at Bird Rocks, the center of the Gulf of St. Lawrence, where Audubon, in June 1833, saw "a strange dimness of the air" caused by the sheer numbers of gannets circling the rock; some odd colonies along the northern fringe of the Gulf; and Baccalieu Island off Newfoundland.

When I first saw the Bonaventure gannetry, it was merely a confusing melange of birds flying, feeding young, playing, or hunting offshore. It took long watching to reveal something of the dynamics of the gannetry. Both breeders and prebreeders occupy the nesting area. The prebreeders are there to experience vicariously the routines of breeding. They can be seen somewhat uncomfortably occupying the fringes of the dense mass of nesting birds. They spend at least the first three years of their lives at sea without coming near a gannetry. During their first year at sea, they probably drive as far south as possible, well into the Gulf of Mexico. But in the next two years they likely become more conservative and may be found off the southeastern coast of the United States.

Once at the gannetry, they are the first to be killed in attacks by dogs or humans. Breeding birds which did not succeed in pairing, or which lost their mates early in the season, join the prebreeders at the periphery of the gannetry and may help to defend them.

A gannetry, though presenting a united front to intruders, is vulnerable. A few years ago John Paget saw an eagle approaching Bonaventure Island's gannet cliffs. He said that he never saw such panic among any living things. "The cliffs erupted," he said. "The birds seemed to go mad. They threw themselves over the edge. I saw many of them colliding in midair. And then, as hard as they could, they all climbed up until they hovered in a big mass over the island. You never heard such a sound of birds crying out. That eagle, he didn't much care for all this and he flew down the coast as fast as he could."

The gannet normally is a standfast at its nest, defending it to the point of death. Fishermen, seeking bait, only have to walk into the gannetry and smash short-handled paddles down on the heads of the nesting birds.

Gulls have worked out at least one method of pirating gannetries. On Bonaventure there are always one, two, sometimes three bright-eyed herring gulls hanging around the gannetry waiting for parties of birdwatchers to make their walk across the island. When humans arrive, the gulls press forward, their bodies intent and eyes watchful. If a human approaches a nest too closely and dislodges a gannet, the gulls dart forward with triumphant cries and seize their eggs or nestlings. It is not hard to imagine dozens, perhaps hundreds, of such gulls in the old days, waiting for the arrival of the then much more plentiful eagles and gyrfalcons.

In the air the gannet is king, but on the ground he's a comic. On foot the gannet waddles, stumbles, lurches. This comic aspect is enhanced by odd individuals which seem prone to sudden and fitful bursts of panic. Once I saw a well-known ornithologist fire a strobe flash at the massed birds at Bonaventure. There was a long pause after the flash, then one bird leaped from its nest and began a half-running, half-flying rush for the edge of the cliffs. She knocked gannets flying in every direction and, at the moment of flight, took more than a score of birds over the edge of the cliff with her. This signaled a general panic. About a thousand birds headed for the cliff edge. Eggs and newly hatched nestlings were swept over the edge. Ornithologically speaking, it was an expensive picture.

Normally, however, a bunch of gannets debouching from a high cliff is an unforgettable sight. They go over the side like paratroopers pouring out of a troop transport plane. Some hurl themselves outward with their wings outthrust and their tails flared; others drop, wings half-closed and tails thrust down sharply like flaps to give them flying control until they build up flying speed.

All bird colonies have their special fascination for watchers, but a gannetry provides material for an anthropomorphic point of view. Gannet ceremonies nearly all have a human quality about them. The first time I saw a gannet *dancing* at Bonaventure, with its feet stamping rhythmically and its neck, wings, and tail moving in a kind of physical counterpoint, I was reminded of an old white-haired man moving energetically through an Irish reel.

A more prosaic explanation, according to the behaviorists, is that gannet ceremonies—among other things—help preserve the "marriage" bond. The ceremonies are diverse and endlessly fascinating. A gannet on the ground can soundlessly direct a flying gannet into a landing, like the batman on an aircraft carrier, by stretching its neck as far upward as possible. The flying bird may be so affected by the signal that he turns down immediately and lands nearby. Both birds greet each other by standing on tiptoe, as it were, beaks thrust upward and clashing together in a moment of intense emotion. The ceremony ends in a flurry of mutual preening.

As the ceremonies acquire meaning for the gannet-watcher, he may be tempted to recall Konrad Lorenz' analogies of the similarities in animal and human behavior. When birds return from the open sea after a few hours of hunting to relieve their mates on the nests, they demonstrate this. The birds change places, sometimes not without a struggle. The female, in particular, may be reluctant to move off her egg. The male may have to hit her firmly with his beak to dislodge her. Once moved, she seems ceremonially bound to welcome her mate's return. She walks some distance from the nest—which may be quite a feat in itself, with every square foot of nesting area fiercely guarded by other gannets—until she finds a bundle of feathers or dried seaweed. She returns, presents her find to her mate. He, however, is not grateful. He seizes the gift and attempts to wrench it away from her. She clings to it and the tug of war goes on until both birds have exhausted their need for ceremony. The seaweed is added to the nesting mound and the female prepares to take off for the open sea.

Sometimes the takeoff is simple, but often it, too, is ceremonialized. The bird seems to wind itself up for flight. It raises its head high, pivots its body back and forth, swivels both eyes *and* sockets inward, like a cross-eyed human. It is a bit of a shock, seeing a gannet in this position for the first time, to realize that it is actually looking down and *under* its beak. If you happen to be close to such a bird and can watch its takeoff path, the effect is grotesque. The bird seems to be having some sort of fit. Suddenly it bounds into the air, thrashes its wings, and is away.

At least, that is the theory of the takeoff ceremony. Often it precedes a minor disaster, particularly on Funk Island, or White and Great Barrier islands in New Zealand, or for that matter on any other gannetry that lies on low flat land. Lacking a cliff to launch themselves off, the gannets hurl themselves into the air and immediately crash, usually knocking down several birds. But once the takeoff procedure has begun, it must continue. The gannets jump again and wings flail all the nearby gannets flat. On a calm day, a takeoff may cut a swathe fifty or sixty feet long before the bird is airborne.

The comedy of the takeoff is a step from the sometimes touching pathos of young gannets leaving the gannetry in the early fall. After about forty days of incubation and about two months of nest care, the young gannets in their mottled brown plumage are ready to fly. They gather in groups, as though to gain communal courage. It has always been my impression that they are deserted by their parents, like young petrels in their burrows, but the ornithological literature does not confirm this view.

Certainly the youngsters are hungry and they look apprehensive. They waddle back and forth at cliff edge. Far below, the hiss of breaking waves hints at the distance of the sea. The young birds, it should be remembered, have never flown before, or even practiced taxiing. They have one chance to fly, and it must be perfect. I watch one bird teetering on the edge. He leans forward, sees the depth of the drop, half falls, and with wings beating regains the cliff top. Another bird stretches his wings out, feels a wind gust pushing him, and launches. It is, in human terms and perhaps for him too, an appalling moment because instead of rising and floating, he begins a long, sideslipping fall toward the sea. Halfway down, and with a hundred feet still to fall, he is out of control, but somehow he slows his ragged flight and, mere feet from the water, pulls into a glide which eventually sends him smashing into a wave. He bobs up unharmed. Some youngsters are not so lucky. They falter at the edge, slip and fall down the cliffs, often breaking their wings or being killed on the rocks below.

The gannet does not deserve to be remembered in moments of such humiliation. Perhaps Yeats was thinking of gannets when he wrote "I would that we were, my beloved, white birds on the foam of the sea!" because no other bird better exemplifies freedom and grace and confidence on the wing. My last view of gannets was at Bonaventure in the late afternoon. A heavy overcast lay over the island. The gannets were almost silent with odd birds returning from the sea hunt.

A roll of thunder suddenly opened the overcast. The sun blazed out. Rain hissed out of a cloudbank. With a concerted roar, thousands of gannets took to the air and opened over the cliffs in a living umbrella. This was no panic. It suggested a joy of life and a harmony of existence that reached and touched the human observer with its perfection.

A BELL FOR RAJAH,
BY JOHN K. TERRES

A vulture is hardly a typical childhood pet, and hardly typical were the lessons learned by an eleven-year-old boy, the grown-ups in his life, and the domestic animals in his neighborhood when he brought home a downy chick found in a rotting woodland log.

I still remember the bright spring day when I found my first turkey vulture nest. I was eleven years old and living in a village in southern New Jersey. I had walked several miles from my home at the edge of town, as I did almost every day, to explore an oak forest, a creek, and a marsh.

As I neared the top of the wooded slope, I saw in a clearing a large hollow log. Suddenly a vulture ran out of it and flapped into the air. I stood startled, not knowing whether to run or stay. I had always been fearful of vultures, and I was relieved when the bird glided away. Scared as I was, however, I had to know what was inside that log. I ran to it, peeped in, then crawled within. In the dim light, on the rotting bottom, I saw two large white eggs.

When I returned a week later, the eggs had hatched and two small downy white vultures sat clumsily on the floor of the log. I crawled to them and picked them up as I would baby chicks in our poultry house at home. I stroked them for a while, running my fingers over their homely heads and sharply hooked bills. Their eyes were open and they made no protest at my handling of them, other than a low hiss. They gave off no odor. The smaller one seemed so helpless that I could not bear to leave him. I left the larger one in the log but took the smaller one home.

That was the way Rajah came into my life. I called him Rajah because, even when he was very young, he held his black-skinned head high and had a defiant way of sitting back proudly on his haunches and hissing fiercely when I tried to pick him up. Later, when I learned Rajah's language—I never heard him make a sound other than a hiss—I knew what he meant simply by noting how softly or loudly he hissed. A low hiss, which he always gave when I was feeding him, meant he was pleased; a loud vigorous one, that he was angry.

Everyone in the village soon knew that I had a vulture. It was unthinkable to many people—*no one* had ever kept a *vulture* as a pet. Men, and some boys of my own age, came to our yard to stare at my big white chick and to remark sarcastically that anyone must be crazy to want a stinking vulture for a pet.

I defended Rajah—he had no odor, at least not at that time. Besides, I had always pitied turkey vultures. Life seemed so difficult for them—they were always gliding about searching for food. It seemed to me that any bird that had to wait for another animal to die might, on occasion, have to wait a long time.

The soaring flight of the turkey vulture, or buzzard as it was called locally, was the most graceful of any bird I had ever seen. Yet everyone in my village spoke of vultures contemptuously, or in disgust—they were scavengers that fed on the putrefying flesh of other animals. Vultures seemed to be doomed as outcasts from other birds, too. They were always on the fringe of bird society, never associating with others, except their own kind.

Farmers and hunters, in their ignorance and prejudice, often shot and killed vultures simply because they thought they were ugly, and because they made easy targets. Naked-headed and hunched, they sat about on fenceposts or in trees not far from the carcass of some animal, awaiting their turn to feed. No one spoke of them as a sanitation corps—that they were performing a useful health service to the community in cleaning up the carcasses of the dead animals and helping to prevent the spread of dreaded anthrax and cholera. My schoolteacher had told me that. I knew also that, someday, these people who hated vultures might even kill Rajah.

But at the moment I had a more urgent worry than the disapproval of the villagers. I did not know what to feed Rajah, or where I might safely keep him. My mother did not want me to have Rajah, and much as she liked birds, she did not want a vulture in the house. I decided to keep him in the poultry yard —if the chickens would accept him.

Rajah was almost as large as some of the hens themselves, and when I first released him, they pecked at him and tried to chase him away. But Rajah hissed loudly, spread his wings, and struck back with his sharply hooked bill. Immediately this established Rajah right at the top of the peck order in the henyard. However, it was to be different when the "silver-lace" Wyandotte rooster came. But that was much later, and the one-eyed monster we called "Long John Silver" had every right by his poultry-yard standards to do what he did to Rajah. Long John had lost his right eye in a fight with another Wyandotte, but that did not stop him from fighting if he was challenged.

Although Rajah did not trust dogs, he tolerated our small water spaniel, whose long silky coat was as black as Rajah's adult feathers would be. But if a strange dog came into the yard, Rajah would hiss loudly, then attack it with a furious flailing of his strong wings, black beak, and claws. Perhaps, in the dim recesses of his vulture mind, he associated the shape or color of certain dogs with that of foxes. Hunters said that if a fox den were located near the nest of a turkey vulture, which is usually on the ground in a woodland, a rock cave, or simply on the floor of the swamp, the foxes might kill the young vultures and eat them, or carry them off to feed them to their own young.

I almost lost my vulture the first week I had him. I tried to feed Rajah raw meat scraps, but he merely sat back and eyed them and made no attempt to pick them up, nor would he take them from my fingers. He also refused to drink milk or water from a bowl I set before him. By the third day he began to crouch flat on the ground in his hunger and weakness. I did not dare tell my mother that Rajah would not eat. That would have been to admit defeat, and she would have made me return Rajah at once to his nesting log.

That first Sunday after I had taken Rajah from the nest, while my mother was at church, my uncle who lived on a farm nearby dropped in to see me. I think he had heard about the vulture. He loved animals and had kept many wild pets of his own. When he saw Rajah his usually laughing face went grave.

"He's starving," he said, and he looked at me angrily.

"I don't know what to feed him," I said helplessly.

"Vultures eat dead animals, don't they?" he asked. I nodded.

"Well, we've got to give him something that a young vulture will eat, and I think I know what it should be."

We went into the house and my uncle opened our old-fashioned refrigerator.

He brought out the remains of a cooked chicken we had for dinner that day. Crumbling the meat into bits with his strong fingers, he mixed it in a bowl with some rich gravy. When he had finished we went into the yard and he set the bowl in front of Rajah. But the vulture only looked up at us helplessly. I was ready to cry. Then I had a sudden thought.

"He's always putting his bill into my partly closed hands," I said. "Mr. Daniels at the grocery store says this might be a sign he wants to eat out of my hand."

My uncle smacked his big palms together as though I had made a great discovery. His blue eyes were warm again. "Boy," he said, "I think you've got it! Go into the house and get me that tin cup you drink out of." I ran and got it and handed it to him. With a blow of his big fist my uncle flattened it. Then he poured the soupy chicken gravy and ground meat into the cup and held it before Rajah. Rajah shoved his bill into it, and when my uncle tilted the cup toward him, Rajah began to feed. His bill and tongue made soft noises as he fed.

I never knew until years later why Rajah had refused raw meat yet was so quick to drink from the tin cup. In the beginning, a young vulture does not eat solid food. It takes soft food, regurgitated by the parent, from inside the old bird's bill. Our flattened cup, a receptacle that resembled the shape of a vulture's bill, had elicited an instant instinctive response from Rajah. I think that my uncle, shrewd in the ways of wild things, knew this, or had guessed it. But he let me believe that he had saved Rajah only with my help and with a big piece of luck from a rabbit's left hind foot that I always carried in my left hind pocket.

After that, feeding Rajah was no problem. Besides meat gravy, he even drank milk and water from the cup. One day he surprised me when he began to feed himself by catching lizards that ran across our yard and big grasshoppers that flew up in front of him from the sandy road by our house. I had always been told that vultures ate only dead animals, but Rajah was proving me wrong about his kind. One hot July day I was horrified to see Rajah pounce on a three-foot-long black snake that glided across our lawn. He swallowed it slowly with convulsive gulps, then looked at me in a way that I recognized as his "hungry look." His appetite was improving.

As Rajah grew stronger, his black feathers began to replace his soft babyish down. Each day he tried to fly. At least he *seemed* to be trying to fly as he stood on our lawn flapping his long black wings. But like a plane revving its motors, Rajah was only warming up for future flights because in his practice he never got off the ground.

It was not until August that he began to lift himself into the air, and with his new ability I began to worry again. Someday I might lose him. Later I learned that it takes about a month for a young turkey vulture to hatch and another two and a half months before it can fly.

Now Rajah's efforts began to show progress. He would run across the lawn beating his wings wildly, and he succeeded several times in bounding a couple of feet into the air. One day he practiced so enthusiastically that he almost came to disaster. Flapping his wings and leaping ahead, he crashed into one of my mother's wooden clothes posts that supported the washline. He collapsed and lay there all in a heap, but he was only stunned by the impact and soon recovered.

Each night, after shutting the chickens in the henhouse, I put Rajah inside the adjoining fenced poultry yard where a night-roving fox or raccoon could not get him. Although he was safe inside the pen, I was afraid that one morning before I got up Rajah might take off and I would never see him again. Had I known of the late-rising habits of wild vultures, perhaps I would not have been so concerned.

Before Rajah could fly, I knew that I had to mark him in some way to I could recognize him if he came back, soaring over our house, a wild free bird. Just a leg band or something of that sort would not do. I thought he should announce his return like the town criers of old. I had an idea.

One day I tucked the big vulture under an arm and carried him to the village harness shop. There, while I held Rajah, the harness-maker carefully fitted Rajah's black-feathered neck with a neatly fitting leather strap. Now all I needed was a bell. Somewhere I had heard of a belled vulture, and the idea appealed to me enormously. Rajah, carrying his swinging neck bell, would be as easy for me to distinguish as a flying falcon that is recognized by her master from the tinkling of her leg bells.

I boarded a country streetcar, and three miles away in a small neighboring city I found what I wanted. It was a small, lightweight dinner bell, displayed in the window of a hardware store. When I held it in my hands and shook it, it had the loveliest sound of any bell I had ever heard.

Back in my yard, I caught Rajah, held him under one arm, and tried to fasten the bell to his leather collar with a piece of wire. Rajah was big and strong, and he squirmed and tried to beat his wings. I had carefully pinned them to his sides, but I had forgotten about his sharp claws. Before I finally fastened the bell to the strap, one of my arms was bleeding from the raking of Rajah's claws in his struggles to get free. But Rajah had his bell, and I knew that if he did fly away, I would be able to recognize him if he ever came back.

One afternoon in September, when I came home from school, Rajah was gone. I didn't know how or when he had gotten into the air, but presumably one of Rajah's takeoffs had been successful. I was sorry I had not been there to see it, and I worried for fear he would never come back.

The next day, when I came home for lunch, I heard Rajah's bell tinkling. He was sitting on one of the tall posts of the poultry yard, and by the way he craned his neck eagerly toward me, I knew he was hungry. He flew to the ground and waddled to the tin cup I held toward him. I had filled it with his favorite meat and gravy. When he came near, however, I got a shock. The odor that came from Rajah was rich, overripe, and nauseating. Somewhere out in the fields or woods he had fed on the dead carcass of some animal, perhaps in company with other vultures.

I held my nose with one hand while Rajah fed from the cup that I held in the other. When he had eaten his fill, Rajah turned away. He ran across the lawn, flapping his wings heavily as he went, then rose into the air. He circled the house once, his bell tinkling faintly, then soared out over a field. As he spiraled upward, he moved away with a strong wind that carried him far out over the pine forest the villagers called "Horse Heaven." It was a remote place where the village sanitation wagon dumped dead horses and other animals to be disposed of by the vultures.

By now I was familiar with an ancient village argument: Did the turkey buzzards that filled the skies over Horse Heaven at the dumping of each newly dead animal find them by their sharp eyesight, or by their sense of smell? There were heated claims by hunters and fishermen who took opposing sides, but I was unable to help settle it. Rajah had never shown me exactly how he knew I had food for him, even when I hid it behind me in the tin cup. If I stood before him without food, he stared at me accusingly out of his black eyes, but if I did have the cup with the mixture in it, even though I kept it hidden under my jacket, Rajah rushed excitedly around me. He *knew* that I had the food hidden on my person. I always suspected that he smelled the concoction, although I was not sure.

Rajah did not come back, and about a week after he had gone we got Long John Silver, the Wyandotte rooster. My father said the rooster would protect hens from attacks by hawks. A few days later, a Cooper's hawk swooped into the poultry yard and tried to seize a hen. Now Long John proved his worth. He leaped high in the air, his spurred feet thrust out at the hawk. Repelled by the fury of the rooster's attack, the hawk flew rapidly away. Now Long John *knew* he was the cock of the walk, and his one-eyed fury was directed even toward me if I came too close to the pen.

It was a week later that Rajah came back. I had eaten my lunch and was out on the lawn when I heard the tinkle of his bell. He soared over the yard, dropping lower and lower. Then he glided down and alighted in the poultry yard. Rajah was used to the hens, but now he got the shock of his young life. At that moment, Long John Silver dashed straight at Rajah. The fury of the rooster's attack almost knocked Rajah over, but he quickly recovered.

I will say this—Rajah was no coward. Although outweighed by five pounds, he raised his six-foot spread of wings and rushed at Long John. He struck at the rooster with his hooked bill, but the eight-pound Wyandotte, besides being heavier, was much too agile for the three-pound vulture.

The rooster clinched savagely with Rajah, pulling out some of his feathers and striking him with spurred feet. The vulture turned away, ran a few steps, and with a silky rustling of his wings lifted into the air. Long John, possibly stunned for a moment by the size of his foe rising above him, paused. In that instant, Rajah left the chicken yard and swung upward and outward, his bell tinkling, his black wings beating heavily on the air. He soared far across a field, turned into the wind, and circled up and up until he was a black speck in the sky. Then he drifted southward and disappeared below the blue horizon.

I never saw Rajah again. After I moved away, some of the hunters and fishermen of the village told my uncle that they heard a bell tinkling in the skies the following spring.

I could not forget Rajah, nor did I want to. He had become a living example of his kind, and of the cruelties inflicted by the blindly prejudiced on the misunderstood. Rajah had been my first wild pet, and he had been the first to teach me that language is not necessary to understanding. Like all wild animals I have ever known, Rajah had a way of communicating that was distinctly and very beautifully all his own.

THE DANCE ON MONKEY MOUNTAIN (AND OTHER CROW DOINGS), BY JOHN MADSON

Who gives the common crow a second glance, unless to condemn it without benefit of trial for alleged crimes against things we hold dear, like crops and ducks? Perhaps we should pay more attention. Not only are the crow's sins greatly exaggerated, but a lifetime afield in the company of these brazen black brigands leads one to the conclusion that the crow is a most remarkable bird indeed, a species that has included in its evolution the most useful features of all its avian brethren.

On my edge of the pavement a road-killed fox was being attended by a pair of crows. There was oncoming traffic and I was unable to swerve, and the wheels of the car passed within inches of the defunct Reynard. The crows calmly gauged my approach as they always do, floated up into the air as I passed under them, and dropped back down to their dinner. They had calculated my brake horsepower and closing speed with the acuity of an Indianapolis pit crew.

They were in character, polishing off an adult red fox that had been unable to coexist with traffic. A red fox is pretty handy at coexisting with almost anything, but he's not as handy at it as the common crow. Few critters are.

In thirty years of tooling along country roads, I have seen many sly creatures that had been clobbered by cars: mink, coyotes, red fox, gray fox, white-tailed deer, bobcats, and even one lightning-rod salesman. Plus a multitude of poor little goofs that never seem to adjust to traffic: a host of songbirds, sparrow hawks, cottontails, ground squirrels, and garter snakes. But never a crow. I've yet to see a crow that had been hit by a car. Crows have a natural immunity to traffic, and experience gives them booster shots. Call it adaptability, or skilled adjustment to familiar feeding situations. But it's more likely a deep and abiding case of the smarts.

If the common crow isn't the endpoint of current avian evolution, he can't be far from it. Sleek, hard-feathered, adaptable, and superbly generalized, he's a bird for all seasons. It's as if the most practical, general features of the class Aves had been built into one bird, plus a low sense of humor and a raffish cleverness that is disturbingly familiar. Most birds behave like birds, endearing themselves to us as a flash of color, a burst of song, or a high, aloof vigilance. But crows may remind us of us—something that's hard to forgive.

I like crows.

Oh, I've hunted them often enough, but this was never attended by any rancor and it always reinforced liking with respect. To hunt crows is part of an acquaintance process, and reveals many facets of crow character. To be mobbed by a raging gang of black banditti teaches one thing; stalking a lone sentinel teaches another. Like men, crows may yield to mob madness and commit insane indiscretions that range from mayhem to heroism, and within the hour they return to being keen, perceptive masters of themselves and their world. The common crow is an animal of many parts that may be more clearly revealed to an ardent predator than a casual observer.

My bond with crows began long ago during certain bitter winter evenings when we traveled together. For several winters at the tail of the Depression I market-hunted rabbits in central Iowa—a grueling enterprise that resulted in a lot more seasoning than revenue. Among other things, it taught me that it's a little easier to be famished and weary if you're not alone. And when I'd turn homeward at day's end and face the miles of crusted snow that lay between me and supper, there were usually a few crows for company—seeming just as tired and hungry as I was. They would escort me in little tattered flocks, beating patiently into the bitter wind, going home. The lights that were beginning to glow in the windows of distant farmhouses were not for us. We belonged to no one but ourselves and to the bleak world whose graying land faded into grayer sky as evening came on, where nothing moved but a few wind-buffeted crows and a boy stumbling through the iron twilight.

Most other birds had gone south; indeed, so had most crows. Those that escorted me were the toughest of a tough breed—friendless and persecuted with every man's hand turned against them, gleaning a hard living from a hard land. Like me, they had spent the day searching for the Main Chance—and like me, they probably hadn't found it. We were hunters and scroungers fallen on hard times together, heading for our home roosts after a day of deficit spending. Those crows didn't have much and neither did I—but for a little while there we had each other to divert our attention and ease the last long miles to home.

When bluebirds and swallows return, I will always welcome them with relief and affection—but never with the respect that I hold for the crows that never went.

Some general comments on a general bird:

Crows are members of the clan *Corvidae*, of course, a tribe that includes all manner of scamps: the jays, jackdaws, rook, magpies, and ravens. The common crow's full name, *Corvus brachyrhynchos*, may not flow off the tongue as do some scientific names, but it somehow fits the owner. Next time you hit your thumb with a hammer, try saying it fast several times. It's good substitute for profanity and doesn't corrupt listening kids.

Spanning twenty-four inches across his slaty wings, with a total body length of about twenty inches, the crow is probably our largest generalized bird. Most other birds his size are somehow specialized in structure and function, but the crow is simply an outsized songbird. His powerful beak is adapted to picking and pecking, and is as useful in predation and carrion-feeding as in field-gleaning. Almost anything that's edible—from seed to grub to nestling—can be crow provender. What with one thing or another, the common crow has the physical and psychological equipment to exploit almost any possibility that comes along.

One day last summer I drove through Greenwich, Connecticut, heading for downtown Greenwich, and while pausing at a stoplight I happened to glance LaGuardia Airport and the flight home. There was a big new office building in up at the roof. Far above the street at the edge of a lofty cornice was perched a lone crow. He was calmly surveying the confusion below, wondering how he could put it to use and looking as self-possessed as if he'd been in a Kansas cottonwood.

I caught my plane and returned to St. Louis, bailed my old truck out of the parking lot, and headed for Illinois. Naturally, I hit Lindbergh Boulevard just in time to be ingested by the four-thirty traffic. And as I inched along, I

157

happened to catch a whirl of motion above a small tree in a nearby factory area. A lone crow was being harassed by a small gray bird, and as they flew over the traffic jam just ahead, I could see that the crow held an egg in its beak. Twice in the same day, a thousand miles apart, I'd seen crows busy being crows in heavily urbanized situations.

It wasn't surprising. I've often seen crows in untidy downtown shopping centers at dawn, checking litter and debris. They commonly haunt manicured suburbs at first light (not usually making much noise about it), and I've seen them strolling across lawns on Chicago's near North Side only a few miles from the Loop. By full daylight, they're gone.

Yet, crows are basically farmland birds with a yen for landscapes that mix trees, cultivated fields, feedlots, and pastures. They are rarely seen in real wilderness, when corvids are likely to be represented by ravens and jays.

Old settlers in the midwestern states claimed that crows were rare on the early prairies, even along the breaks of the Missouri River. Crows apparently arrived in eastern Nebraska in the early 1860s, exploiting changes that were beginning to evict the indigenous magpies and ravens. The common crow flourished with the advent of prairie farming and the planting of windbreaks and tree claims; in the East, much the same effect was achieved as the vast expanses of original forest were opened by settlers.

If it's true that there are more crows today than at the time of Columbus, then the golden age of crowdom must have been in the waning years of the nineteenth century when American farmland as a whole possessed the somewhat ragged quality that Crow seems to dote on. It was a time of small family farms, of diverse grain and livestock production, with young tree claims, maturing on the prairies and remnants of original forest still surviving in the East.

Crow moved into the new situations with the raffish abandon of a pickpocket at a Republican convention. Man cleared and toiled; Crow jeered and foiled. The black brigands committed outrageous acts of disrespect and got away with it, moving the Reverend Henry Ward Beecher to reflect: "If men had wings and bore black feathers, few of them would be clever enough to be crows." Their brazen affronts to the American farmer are pretty well typified by the two Florida crows that were once seen perched on a cow's back. It was February, when Florida crows are scrounging nest materials, and each of the birds had a beakful of white hairs that had been plucked from the back of the cow.

That cow probably wasn't flattered by the crows' attention, but it could have been worse. In areas of crow concentrations in Kansas, there are reports from angry farmers that newborn calves are being blinded and even killed by crows. At the same time, there are complaints of heavy damage to milo and winter wheat.

Some depredations of crows, however, may be more apparent than real. Although corn is claimed to be the favorite food of crows, a study conducted in five New York counties indicated that corn was less than fourteen percent of the crow's annual food—and most of that was taken during winter. In May, when corn is sown in central New York, it amounted to only one percent of the diet of crows studied. Other work has shown that confined crows preferred live mealworms to all other foods offered (including grain), which supports one biologist's findings that a single family of crows may account for forty thousand grubs, caterpillars, army worms, and other insects during the nesting season alone.

In my field of game management, Crow has long been regarded as a prime spoiler—especially of waterfowl and pheasants.

There's no doubt that crows will watch the movements of adult ducks, and if an incubating duck is flushed within sight of crows, they can easily find her eggs. Even though the duck covers eggs before leaving her nest to feed, this may not help if crows see her take off. It has been noted that duck nests with good concealment (by human standards) may be destroyed just as readily as exposed ones—in fact, the better-hidden nests may even suffer the most crow damage. On one marsh area, only four ducklings hatched from two hundred eggs. Crows got the rest. Of over five hundred duck nests once studied in the prairie provinces of Canada, about half were destroyed before hatching. The crow led the list of predators, taking thirty-one percent of the nests.

But in spite of such things, most game managers feel that crow predation is only the final stroke in a series of events that doomed eggs and ducklings from the start.

During his classic study of the blue-winged teal, Dr. Logan Bennett found "pecks of crow-destroyed [duck] eggs" around Iowa marshes—but noted that practically all of these had been promiscuously dropped before serious nesting had begun. They did not represent destroyed nests, nor eggs that would have hatched. He also found that heavy predation on ducklings was most likely to occur during drought periods, when large numbers of young ducks were confined to shrinking water areas.

A ring-necked pheasant nest that is destroyed by crows is likely to be one that had problems from the beginning. For example, a nest that's barely hidden in a meager strip of fenceline grass between two open fields. Or a hayfield nest that is discovered in time by a farmer who mows around it and leaves it undamaged in a tiny island of unmown clover. To a passing crow, that little deviation from the norm is intensely interesting. And once he learns that such a place may hold pheasant eggs, he never forgets.

The furies of Hell are transcended not by woman scorned, but by Crow enraged.

A sample of that superfury was once loosed on my old friend Bruce Stiles, late director of the Iowa Conservation Commission, when he was a young game officer stationed along the Missouri River in the early 1930s. Late one day during the waterfowl season he was walking alone across one of the Missouri's vast sandbars when he found a crippled crow. There were geese in the area, and Bruce was reluctant to fire a shot, so he tried to dispatch the crow with a pole cut from a sandbar willow. However, the makeshift club proved too light and resilient to do the job. And as Bruce chased the fluttering bird across the open sand, lashing at it with the limber pole, the crow set up a clamor of pain and alarm.

It was the time of day when far-foraging crows are beginning to converge on the great river roosts, and may loaf on the river's immense sandbars before flying to their roosts. Almost at once, angry crows arrived—materializing out of nowhere in the way crows will in such situations. The first shock troops quickly grew into a great mass of birds, their rage and frenzy intensified by the steady arrival of reinforcements. Within minutes, Bruce was the nucleus of a raging horde of crows. He never knew if there were hundreds or thousands; it was simply a roaring black cloud that engulfed him. He was being struck about the head and shoulders, and as he averted his face and shielded his eyes from attack, a striking bill laid open his unguarded cheek. Bruce was carrying a seven-shot 12-gauge gun, which was then lawful for waterfowling, and emptied the full magazine into the mass of birds. He loaded again, trying to protect head and face as he did so, and triggered another seven-shot volley. A 12-bore shotgun firing express loads at close range is a thunderous weapon, but the second volley had no more effect than the first—if anything, it only heightened his attackers' fury. Bruce beat a fighting retreat to the shelter of a distant willow thicket, and only then did the crows begin to draw away.

I can half-remember the old tales of hunters who spoke with respect and wonder of the mob frenzy of embattled crows, warning of the risks that a lone gunner might face. I always discounted such yarns, regarding them as rural precursors of Daphne du Maurier's "The Birds." But since then I have known Bruce Stiles—and have often hunted crows over owl decoys.

For pure, unalloyed hatred, nothing in nature can compare to Crow's attitude toward the large owls. Housecats and hawks can incite crows, but their fullest frenzy is reserved for their living nightmare.

To Crow, a large owl is every dark and fearful dream come true. All of Crow's wit and wisdom is to no avail against the onslaught of Owl, and all crows know it. In the black midwatches of the night, a great horned owl will sweep through a roost like the Angel of Death, soft and silent and consummately deadly. Nor is it just a matter of one owl seizing one crow; the owl may strike repeatedly, feasting only on heads and brains. It is a nightmare that Crow remembers through all the daytimes of his life, regarding owls with a primal dread and hatred that most men have happily forgotten.

When Crow discovers a great horned owl during the security of daylight, there is an instant clamor. From near and far, shrieking crows rally to the special battle cry that seems reserved for such times, crowding around the owl in a whirl of outrage. In full light they are more than a match for their enemy, and they press the advantage as Owl glowers at their insults, strangely reluctant to fight back. There is a running debate about whether crows actually strike an owl, but I have crept to within easy binocular range and watched crows strike feathers from the back of a wild, unrestrained great horned owl. (But never from the front, with those baleful eyes and terrible feet.) Crows could surely destroy an owl if they were willing to pay the price, but I've never heard of them doing to. Not a living owl, anyway.

When one of my classmates in graduate college was doing pheasant research in northern Iowa, he sometimes managed to work in a bit of crow hunting. Early one fall he borrowed a mounted snowy owl from his major professor's collection, attached the owl to a tall pole, and erected it in a wood-lot on his study area. The results were something less than spectacular; most of the morning he never fired a shot. Come noon, he left the decoy in place and joined his farmer-cooperator at lunch. Five cups of good Norwegian coffee later, Chris returned to his blind. A band of crows was just retiring with victorious jeers, the decoy had vanished, and the woodlot was white with a blizzard of owl feathers.

Winter coming on. Time for the Grand Reunion.

Some northern crows never do migrate, but the typical crow is as migratory as any other passerine bird and undertakes a journey that is as prompt and methodical as a robin's. It's not usually far. Few crows migrate more then five hundred miles—going just far enough to assure themselves of food supplies that won't be locked up by prolonged snow cover.

Most of the crows reared in the north tend to winter in our mid-latitudes between the 40th and 35th parallels. At the eastern end of this band there is a great center of wintering crow populations in the vicinity of Chesapeake Bay and its tributaries. One of the midwestern centers is near the junction of the Ohio and Mississippi rivers. A little further west, large numbers of crows winter along the Arkansas and Missouri rivers. From early December into February, immense populations of crows concentrate in the regions that lie just south of winter and just north of spring. In 1886, ornithologist Samuel Rhoads said: "In winter a radial sweep of 100 miles, described from the city of Philadelphia and touching the cities of New York, Harrisburg, and Baltimore, will include in the daytime in its western semicircle fully two-thirds of the crows inhabiting North America, and at night an equal proportion in its eastern half." Which wasn't true, of course. Similar claims could have been made all the way out to Kansas City, and beyond. The part to believe isn't Rhoads' flawed conclusion but the impression that inspired it, for the daily flights of crows to and from their great winter roosts might lead a man to say almost anything.

Such roosts may be small in area, but huge in terms of occupancy. In 1886 and 1887, up to 200,000 crows occupied twenty acres in Arlington National Cemetery. There was a twenty-five-acre grove in New Jersey near Hainesport that held as many as 300,000 birds. In Pennsylvania, the twenty-acre Davis Grove in Montgomery County had over 200,000 crows. The immense winter roosts in the vicinity of Chesapeake Bay caught the attention of early naturalists, and both Wilson and Audubon wrote graphic accounts of some of the great crow roosts they had visited there.

The general localities of some roosts have been used for centuries, and individual roosts may extend beyond the memories of the oldest men. The main criterion, of course, is shelter. Roosts may be in thick conifers—not necessarily large trees, but dense enough to give good protection during storm periods. Or the roosts may be in deciduous groves, and on some islands in the Delaware River the roosts were in reeds, coarse patches of grass, and low brushwood. The big Arsenal Island roost in the Mississippi River at St. Louis provided trees for roosting in ordinary weather, and during severe weather and storm periods the crows often spent the night on the island's snow-covered sandbars or even on the ice shelves that surrounded it.

The greatest crow concentrations today are surely in prairie regions where gun pressure is light, human population is thin, and wintering conditions are ideal. There is a big catalpa grove in central Kansas today that is said to harbor ten million crows in midwinter. State Forestry, Fish, and Game Director Richard

Wettersten found this hard to believe—until he saw it. "I don't know how many there are," Dick told me recently. "But when you get into a roost that size a few zeros more or less don't have much meaning. It's beyond comprehension, no matter how you figure it."

The biggest crow roost I've ever seen was on the north side of Lake Fort Cobb, about fifty miles southwest of Oklahoma City. For years, crows had wintered there on a south-facing slope that was densely grown with jack oaks, few of which were much over twenty feet high. Relatively safe (it was on state parklands), this roost was somewhat sheltered from northern wind and was in a region where peanuts and other crops were generally accessible all winter.

We never knew how many crows there were in that roost. It was commonly held that there were at least eight million birds during the roost's peak occupancy during January, but Karl Jacobs, then game chief of the Oklahoma Department of Wildlife Conservation, flatly denied that. "It's a ridiculous exaggeration," Dutch used to say, "and I doubt if there have ever been more than four million crows in that roost!"

How many were there, really? Pick a figure. My conservative training used to limit me to a million, but you could quadruple that and no one would argue. I only know that when I first saw it, it was one of those crowd phenomena that rank with July in Disneyland.

We would stay at the Lakeside Motel just across the sandy road from the main roost, and sometimes when there was no moon I would sneak over into the roost to hear the universal mutter and babble that went on all night. From a short distance away it was like the sound of an ocean at flood tide, or night wind blowing through a forest of pliant trees—the ceaseless murmur of an indescribable multitude of large birds talking in their sleep. From closer in, it would resolve to low warbles that were strangely robinlike, or muted henlike cluckings and a wild assortment of grumblings and night-mewings, but rarely any *caw* fragments or anything that resembled conventional crow noises. The crow has seven pairs of syringeal muscles that give it a wide range of vocalization—and the subtle undertones of that range can be heard only in a crowded roost on a moonless winter night. It's been said that Satan never really sleeps. Nor do his imps, apparently.

With the approach of day, in the limbo of light and darkness that my Ozark friends say is "before the crow and after the owl," the muttering would begin to take on definition. The roost was awakening.

For about an hour the crows stayed put, shifting about and expressing themselves in a rising babble. It was as if they were staying in bed to organize their thoughts about the day's work, and in no particular hurry to be up and at it. If a few crows flew up out of the trees, the main host would shout angrily, telling them to get the hell back in bed and stop bucking for promotion. Finally, however, several thousand crows would take wing and begin milling above the main roost, yelling down at their companions and making wild threats and promises. This would ignite a wild enthusiasm that could be heard several miles away on a quiet morning, and an immense canopy of crows would rise out of the trees and begin streaming out to the day's feeding points.

They dispersed widely, often thirty miles or more, hunting that Main Chance. During the day their numbers were unimpressive, although we could always see a few crows feeding in fields, and bits of black embroidery in the high Oklahoma sky.

But by late afternoon, the day's work nearly over, they would begin staging in hedgerows, creek bottoms, and woodlots, gathering by tens and hundreds for the return home. Their great numbers were again becoming apparent. This was not as obvious if there was a strong wind that kept the return flights close to the ground. But on quiet days the crows would come streaming in high, the flocks coalescing into clouds, and the clouds stretching in unbroken streams of black freebooters that sometimes spanned the horizons, growing ever denser as the dark legions converged on the main roost. On such days when they approached the roost area from on high, they might plunge almost straight downward for a thousand feet, and then flash falconlike with half-closed wings over treetops and fields. They would rarely go directly to the roost during clear, quiet evenings, but loafed in adjacent fields between sunset and bedtime, not settling into the roost until full dark.

We've often wondered, watching a settling-in, if there is an order of seniority. Crows roosting in lower parts of the trees are often whitewashed by morning; can this be a sign of social ordering? And are the dominant crows the whitewashers or the whitewashees? As a whitewashee of some experience, I can't say much for being on the receiving end. But I can't say much for roosting in a treetop, either.

We were usually at Fort Cobb to hunt crows, which is not to say that we hunted the roost. This is simply never done by bonafide crow hunters, who prefer to work miles from a roost with mouth-blown call and camouflage, meeting crows on ground where both hunter and bird must be at their best. Our gunning was done far out along the flyways during the day, and never in or near the main concentration.

I can remember unproductive days when ten hours of effort earned us little more than the sand in our eyes and the growl in our bellies, and other days when we grew slug-nutty from gun recoil. But for some reason, more than anything else, I recall the Dance on Monkey Mountain.

There is a singular mound rising out of the flat sandy country north of Fort Cobb. We never knew why it was called Monkey Mountain, nor did any of the locals, but it was obviously a check point for the morning crow run.

A friend and I were lurking in the Mailbox Blind early one day, not far away. There was a bitter, gritty wind gusting out of the Panhandle, sweeping up the flanks of Monkey Mountain as a flight of late-rising crows arrived. They appeared over the distant mound dipping and rising, hanging above the summit with a wild display of aerobatics, bouncing and rolling in the wind currents and sometimes seeming to fly backward. This first bunch of hot pilots was joined by others until there were probably fifty crows up there, dancing in a stacked aerial ballet that must have lasted ten minutes. We'd never seen anything like it, and it puzzled us until my companion exclaimed: "Why, I'll be damned if they don't seem to be enjoying themselves!"

163

It must have been crazy up there, some erratic and violent combination of wind currents above the summit. Yet, we were sure that those crows were not inextricably locked in currents from which they couldn't escape. They were handling themselves with confidence and a great deal of inventive skill, master aerialists at play, wrestling with the wild wind out of sheer exuberance, like an airlines captain doing aerobatics in a vintage biplane on his day off.

Inevitably, a great roost draws the attention of men embittered by the real or imagined wrongs done them by crows.

Most often, such attention is expressed with shotguns as landowners work over the big roosts in an effort to kill as many as possible and drive the rest away. If they make a steady job of it, they'll probably disperse the crows and break up the main roost into a number of smaller roosts. In terms of actual decimation, it has practically no effect.

During the 1920s and 1930s, however, an efficient method of crow-killing was devised in large roosts.

At 4 A.M. on March 9, 1938, a crow roost near Sharon, Wisconsin, was dynamited, killing "well over 5,000 crows." But this was admitted to be of "small account" when compared to the great bombings in more southerly roosts.

There was a time when Oklahoma roosts were regularly bombed in an effort, ill-advised or not, to control crop depredation. An Oklahoma crow bomb consisted of a crude sheet-iron tube about twelve inches long and three inches in diameter, with a wire hook at the open end. A stick of dynamite was inserted in this tube, and a couple of pounds of iron pellets poured in around the explosive. During the day, when the crows were gone, trees in the roost were festooned with bombs hung at all elevations; they were wired in series and detonated electrically.

The blast came late at night when the roost was filled with sleeping crows. Sometimes, as in the Wisconsin shot, the circuit was closed just before dawn—and perhaps on a Friday night or early Saturday morning so that schoolchildren would be available to gather dead crows. Through all the annals of man's relentless persecution of wildlife, nothing can compare to the instantaneous carnage that occurred when a roaring storm of shrapnel and a wave of concussion swept through a crowded crow roost. Near Harrisburg, Illinois, a thousand steel cannisters containing dynamite and scrap iron were hung in a roost and detonated at one time, and next morning one hundred thousand dead crows were picked up.

We haven't heard of a great roost-bombing for years, which isn't to say that roosts aren't still being destroyed.

Our friend Floyd Kringer, an Illinois game biologist, believes that his crow populations are thinner than they were a few years ago—and he knows why.

Back in 1951, Floyd and a friend checked the crow populations along the Kaskaskia River bottoms in southwestern Illinois. On one route they made twenty-one stops and had "canopies of birds" over them at nineteen of those stops. But that was when the Kaskaskia bottoms were prime country. For example, there was one farmer who controlled a solid square mile of big timber in the bottoms—a tangled maze of little backwater sloughs and oxbow ponds, with broad flats of alluvial timber with huge burr oaks and soft maples. Today, that entire six hundred forty acres of wild floodplain timber is gone, replaced by corn and soybeans. Floyd has no doubt that the local crow decline wasn't caused by shooting (he doesn't know of a single dedicated crow hunter in his territory today) but by habitat losses that have destroyed prime roosting and nest sites. There are still crows along the Kaskaskia, and doubtless always will be, but not in the numbers that were met there before the stream channelizers began making the world safe for soybeans.

Are there fewer crows nationally than there were twenty years ago? Perhaps there are—although U.S. Fish and Wildlife Service reports don't tend to support this. The breeding bird survey that is conducted each June in the United States and parts of Canada indicates no significant change in the continental crow population from 1966 to 1974. There appeared to be a slight population upturn in the East and a slight downturn in the West during this period, but each balanced the other out. When a statistical regression is plotted for that eight-year period, the curve is flat—indicating no significant population change one way or the other.

If these surveys are valid, is our impression of fewer crows invalid? Not necessarily. The apparent decline of crows in some areas could reflect a fragmentation of crow concentration as a result of roost destruction, and a wide-scale reduction in the sizes of crow flocks.

At any rate, recent federal regulations governing crow hunting don't actually reflect a decrease in the national crow population. When the U.S.–Mexico Migratory Bird Treaty was revised in 1973, Mexico insisted on specifically including *Corvidae* in an effort to protect certain jays. Crows were part of the package. As a result, federal law now limits crow hunting in the forty-eight contiguous states to a season not to exceed one hundred twenty-four days each year. The states are given an option to set season dates, limits, shooting hours, methods of take, or whatever—but crows may not be sport-hunted during peak nesting periods or taken by any means except firearms, archery, or falconry. However, none of this applies to crow control in cases of actual or impending depredation, or when crow concentrations may endanger public health. In an odd reversal of priorities, roost-bombing is not outlawed by the new regulations—which curtail the use of a shotgun and a dozen decoys but freely permit dynamite cannisters to be hung in the nesting groves and winter roosts of a protected migratory bird. It's all very confusing, but Crow will probably figure it out before we do.

It is likely, and appropriate, that a coyote will use the bones of the last man as a scent post. Beyond that, it's just as likely that the bones of the last coyote will be picked clean by Crow. If any critter was designed to endure and ultimately prevail, it is Crow. Brigand and buffoon, forever adjusting and adapting and cocking a suspicious eye at the situation, he'll hang in there from sheer perversity, his own bird to the last. As poet Ted Hughes wrote: "Crow—flying the black flag of himself." And at the end, when Crow follows the long procession of species out of a world grown cold under its dying sun, he'll exit laughing.

IN PRAISE OF SNAKES, BY ARCHIE CARR *Efforts to save animals of all kinds from threatened extinction—the peregrine falcon, Kirtland's warbler, humpback whale, black-footed ferret, Atlantic salmon, desert pupfish, monarch butterfly, and a host of other increasingly rare species—have received wide publicity and broad public support. Even such commonplace and sometimes pestiferous species as blackbirds and pigeons have their staunch defenders. But when did you last hear a good word about snakes? They are, after all, the fascinating and useful product of millions of years of evolution.*

When Rachel Carson chose the name *Silent Spring* for her epoch-making book, the silence she had in mind was lost birdsong on a poisoned Earth. The book was a powerful document and people took heed of it, and birds became not just objects of concern but a symbol of our own predicament. This was good for the birds, but it left unattended a lot of other creatures that had no songs to start with, and had been silent all the time.

Snakes, for instance. I want to speak in behalf of snakes. I live in snake country, and have always liked snakes and have kept them steadily on my mind. There is a dearth of good census data to prove it, but snakes seem to me to be disappearing very fast. Their survival problems are much worse than those that birds have. Besides the waning of their food supply as rodents and frogs grow fewer, and besides the growing toll that cars take on multiplying highways, snakes face the indifference or active antipathy of most of the human race.

People who resent snakes explain their attitude by saying that some snakes are venomous. But there is more to the hangup than that, a lot more. For example, if you pull an angry wildcat out of a bag the reaction of witnesses is no more than simple alarm, or perhaps even grudging admiration. But take out a harmless snake, and the faces of the people suggest that you have somehow flouted common decency and the will of God.

I once put into print the risky proposition that a lingering uneasiness over snakes may have come to us genetically from forebears prone to be eaten by pythons and the like. I don't fight for the notion, but I can't logically discard it. In any case those old days when a proper fear of snakes was the mark of the successful ape are long gone, and the remaining vestige, if it exists, is just a useless atavism, and people ought to dominate it, as they dominate the dread of high places or of eating octopus. Above all, we ought to keep silly people from conditioning it back into our young by shrieking when a snake shows up.

I am not suggesting that anybody take asps playfully to their bosom, or wrestle anacondas, or even walk heedlessly through a palmetto flatwoods in Florida. But the United States is blessed with a great diversity of harmless snakes, the numerous members of the huge family Colubridae that do no damage to anyone. Actually, the negative ring of the word *harmless* is inappropriate for these creatures because they have far more positive virtues. Besides being ecologically useful, they are artistically decorated, richly colored, and consummately athletic beings that really ought to be appreciated more. They ought to be watched. People watch birds regularly nowadays, but you rarely come upon anybody watching snakes. That is a great shame, because snake-watching, though a more demanding exercise than birdwatching, can be every bit as rewarding.

Snakes are harder to find than birds. They don't fly about, or sing, or do a lot of overt things that help the would-be watcher find them. And even after you have located a snake, purged yourself of any remaining scrap of prejudice, and made ready to witness engrossing natural history, it may be quite a while before the snake does anything. Birds of course fidget constantly, except perhaps for owls and fishing herons. There is never a dull moment in their company. Snakes, on the other hand, being cold-blooded, live more deliberately; and if you have the misfortune to come upon one that has just eaten, there may be little or no action for quite a while—for up to a week, perhaps, or in extreme cases even longer. Some people don't have that kind of time to put into the venture, and go back to watching birds.

But while snakes are slow in digesting, they are also often slow in coming upon anything to eat, and spend a great deal of time in their foraging. It is then that you see them do exciting things.

To court the quiet pleasures of snake-watching, go out to the edge of town and walk along a grown-up fencerow till you come upon a racer, say—a blue racer if it's the North you are in, or a black snake if you are in a southern state. The racers are subspecies of *Coluber constrictor*. Despite their specific name, they don't constrict, but they do a great many other interesting things and are worth some patient attention. Finding one of them on the sunny side of a fencerow on a fair May morning is no problem. The trick is to keep him in view without breaking into his natural routine. If he sees you first he will either bolt and disappear, or rear back, coil, and rattle the leaves with the vibrating tip of his tail, and then quickly retreat if he fails to scare you. But once your snake-watcher's eye has been sharpened you will begin seeing racers before they know you are there, and then you will be able to watch one thread his way silently through the leaf drift, easing along in slow undulations or creeping on tilting belly-scales as he explores with his flicking tongue-tips every old log or pile of debris where a mouse or frog or smaller reptile might be hiding. If he comes to a pond he will prospect carefully around the edges, moving into and out of the shallows, swimming the small embayments, sliding up silently into low willows to see what bird nests or treefrogs may be up there. To keep contact with a blue racer through a morning's foraging is a challenge that will call out all one's self-discipline. Bird glasses can be helpful, but the main requirement is a serpentine sort of stealth.

A birdwatcher is helped by the singing of the objects of his search. Snakes never sing. Some rural Nicaraguans say that the fer-de-lance is able to whistle, but they are mistaken. Rattlesnakes of course are commonly heard before they are seen, and sometimes one of them will jump the gun and reveal himself by rattling at a man or dog from a long way off. These are probably just high-strung individuals, however, because most rattlers stay quiet until they feel they are about to be stepped on. Another snake that is sometimes heard before being seen is the pine snake or gopher snake (genus *Pituophis*), which rears up in high coils in an impressive way when disturbed, and reinforces the menace by a hissing that must be the ultimate in serpent vocalization. The sound is so loud you can hear it across an acre of ground, and so stertorous that one subspecies evoked the name bullsnake from the early settlers. Pine snakes are active, intelligent animals, really, and quite affable once they shed their racial prejudices; but when first encountered in the woods the show they put on is unnerving, and they sometimes begin their bellowing at you from quite a distance.

Often a mixed chorus of indignant-sounding birds—buzzing wrens, shrieking jays, and chipping cardinals, for instance—will lead you to a snake. Another sign of a snake is a shrieking frog. Some of the most arresting snake behavior is related to their feeding. Watching this is sometimes not for the overly impressionable person, simply because the diets of snakes tend to be heavy in whole, live vertebrate animals. So if you take the Second Grade out on a field trip and come upon a milk snake engulfing a mouse or a bullfrog screaming heartrendingly because a garter snake has hold of his leg, some calming apologia may be in order. You might explain that snakes—having no hands, only little prickly teeth, very distensible jaws, and a liking for extremely fresh food—eat mainly whole, live animals. It is simply their custom, you can say, and not much worse— is it, really?—than a man eating a raw oyster or dropping a lobster into a pot of hot water. In any case, a shrieking frog is an almost sure sign that a snake is there, and you only have to make a quiet approach to be able to get close to him.

A more ingratiating bit of snake behavior to watch is the bluffing and playing dead of the hognose snake, or spreading adder. Most people don't live in spreading-adder country, but those who do are neglecting a local asset if they fail to make an effort to see the antics of one of these short, corpulent, and inherently amiable serpents when he tries to discourage interlopers by pretending to be first a deadly viper of some kind and then the mere corpse of a snake.

Most people who come upon hognose snakes either conclude that they are venomous and dispatch them, or go quickly away. To see their act at its best you should do neither, but behave as follows: First of all, make sure it is indeed a hognose snake you have at hand and not a rattler, which would prove to have almost none of the spreading adder's winning ways. Then, move straight up to the snake and sit down on the ground in front of him. He will coil in a purposeful way, rear back and spread the whole first third of his body as thin as your belt, and lunge out at you repeatedly, each time hissing with almost intolerable menace. If instead of recoiling you steel yourself and reach over and pat the snake on the back, his menace will wilt before your eyes, and he will proceed to prove that you have killed him. He will turn over onto his back, open his mouth, extrude his tongue and rectum, and then, after writhing about until his moist parts are all coated with debris, lie there belly-up as clearly defunct as any snake could be.

But don't feel badly about him. Give him two minutes, say, and the catalepsy will wane. He will draw his tongue back in and ever so slowly turn and raise his head to see whether you are still there. Move your hand quickly before him, and he will flip back over into his supine seizure. Reach down and turn him right side up, and he will instantly twist over onto his back again. But then get up and move off a little way and wait patiently behind a tree, and you can watch him slowly come back to life, turn right side up, and quietly ease away.

Another overt piece of snake behavior that once in a blue moon rewards a lucky snake-watcher is the combat dance that the males of some species engage in. This is a stereotyped, balletlike posturing by two male snakes, a crossing and recrossing of necks or intertwining of bodies that takes place as the two performers raise the foreparts of their bodies high above the ground until they topple over backwards. The routine is punctuated by periods in which the snakes chase each other about at terrific speed. Then the neck-crossing is resumed. No biting or constricting or other violence is involved, and the social function of the dance is not clear. Over and above its interest as an arresting instinctive behavior pattern, the occurrence of the ritual in such distantly related animals as pit vipers, rat snakes, and racers is a thing to wonder about. Either the dance has come down to them from a very distant common ancestor, or it has been hit upon independently by all these different modern snakes. Either explanation would seem pretty extraordinary. As simply a thing to see, however, the dance is one of the rare rewards of the fortunate snake-watcher.

If an aspirant snake-watcher has trouble getting out into open country he can find a lot of gratification watching well-adjusted captive snakes. Not captive in the sense of caged; but domesticated, living unrestrained in the house with the dog and children. Every kind of snake is not appropriate for this kind of relationship. Water snakes, for example, get badly underfoot and may bite when you trip over them; racers upset bottles in cupboards; rat snakes insinuate themselves into all sorts of small cracks and secret places, and often get mashed in doorjambs. None of these are insurmountable objections, but the point is that there are better snakes for the purpose; and the best of all, in my opinion, is the indigo snake, *Drymarchon corais*. The indigo is handsome and extroverted, and for some ironic reason a natural-born friend to man. It rarely bites even the first hand that lays hold of it in the wild, and quickly adjusts to life in houses with people. Not being a constrictor, it never creeps surreptitiously about the walls and woodwork, and being heavier and less nervous than the racers, it is not so inclined to whip petulantly about among human legs in a room.

Actually the indigo snake is becoming dangerously rare throughout most of its range, and a person probably ought to go to jail for catching one. Indigos can be bred in captivity, however, and it really ought to be done, because a more ingratiating house pet is hard to imagine. I have known several people who have shared their homes with indigo snakes. I knew a mestizo family in Honduras who did so. Theirs was not a true indigo but the brown subspecies that occurs in Central America, and has much the same personality as its burnished-blue northern relative. That snake simply took up in the Saenzes' house and was allowed to stay on because it was thought to keep rats away. It lived in a loftlike attic where corn was kept and came down through the living quarters when it wanted a drink of water from a tank by the kitchen door.

My friends Bill and Joan Partington had a rewarding relationship with an indigo snake that lived in their home for many years. It stayed so long that Joan had a baby while it was there. When she nursed the child the snake would come out of the bookcase it lived in and rest the forepart of its body on the arm of her chair to watch the undertaking. The Partingtons could never be sure whether it was the baby that evoked the rapt attention, the way Joan was built, or both. Bill even suggested that the snake might have thought the child was trying to swallow Joan, but that seems unlikely to me.

But actually, while keeping snakes in homes and zoos is fun, it does not discharge our obligation to save them as wild species in the world around us. Most sensible people have now worked up a healthy fright over the future the world holds for their children, and birds are often drawn in under the umbrella of their concern. But I have heard little worrying over the future of snakes, and this to me is depressing. Snakes are not degenerate beings, punished with leglessness for ancient sins, as people once said. A snake is the elegant product of a hundred million years of natural selection. Its loss of legs was an evolutionary advance, a means of living successfully in unexploited ways. But because those ways are secret, the decline of snakes in our changing world has gone on almost unmonitored. It is several years now since I last saw an indigo snake cross the dirt road I drive home on, and I take this as a straw in the burdened wind of the times. Others have spoken for cranes and whales, and I hasten to say these words in praise of snakes, whose silent spring is also far along.

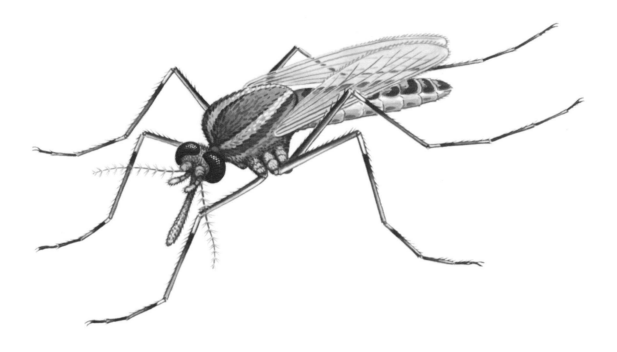

IMMORTALITY,
BY C. BROOKE WORTH

Although a heretofore unknown bird may be discovered every few years, we can state with reasonable accuracy that there are only about 8,600 species of birds on Earth. In the world of insects, however, no such neat number exists. Upward of 900,000 species already have been formally named, thousands of new insects are added to the list every year, and no entomologist wants to guess how many hundreds of thousands (or millions!) of insects still await discovery. But despite such unfathomable numbers, a small measure of fame accrues to the hard-working scientist whose own name comes to be appended to the official description of, let us say, a mosquito.

I knew nothing about African mosquitoes when I first arrived in Natal province in 1958. I was assigned by the Rockefeller Foundation as entomologist to the Arthropod-Borne Virus Research Unit (ABVRU) of the South African Institute for Medical Research, in Johannesburg. My job was to identify mosquitoes caught at our field station in the Ndumu Game Reserve, in the lowlands near the east coat. We shipped the mosquitoes, frozen in tubes in dry-ice containers, by plane to Johannesburg. There they were thawed, ground up, and inoculated into baby mice. If viruses were present, they would reveal themselves by causing illness or death.

The commonest mosquito at Ndumu was one called *Aedes circumluteolus*. Soon, with the help of three South African colleagues in Johannesburg and an excellent mosquito book written by F. W. Edwards of the British Museum (Natural History), I was set along the path to becoming an expert.

Well, after you have examined 81,702 *A. circumluteolus*, as I did during two years' time, you *ought* to be getting down to fine points. And that is how I became immortal. On a January day in 1959 I was sorting mosquitoes as usual when I came across a specimen of *A. circumluteolus* that looked a bit odd. I wiped the tube carefully, saw to it that my glasses and my 5x eyepiece magnifier were clean, moved to the window for maximum illumination, and looked again.

The mosquito was definitely "different," though at a quick glance it could have passed for *A. circumluteolus*. What drew my attention in the first place had been the uniformly yellow, *unstriped* dorsum of the thorax, though that might still be a rarely permissible variation in a somewhat unorthodox member of the species. But at the limits of vision with my weak lens, I thought I could see that the wing-veins bore all sorts of yellow scales that far exceeded even marginal propriety.

I set the tube aside for further study later, because now my job was to get through with the rest of the day's catch so that I could put my native assistant, Qmba Ngwenya, to work with the killing jars. But only a few specimens farther on I came upon a second aberrant *A. circumluteolus*, if that is what they were. This was ridiculous! How could two look-alike freaks crop up on the same day? Despite the urgency of my work, I took enough time to compare the tubes side by side. There could be no doubt of the mosquitoes' resemblance in every detail.

In all cases of questionable identification, or if I knew that a particular specimen was rare or was perhaps needed in our collection in Johannesburg to complete an adequate series, I made pinned mounts of the mosquitoes, labeled them, and stored them in special insect boxes where they would be safe from damage. That happened so infrequently that it made no lamentable dent in the numbers of specimens submitted to the lab for virus studies. Indeed, it was part of the virus plan anyhow, for once the problems that such mosquitoes might present at first were solved, future captures of the same kind could be processed confidently in the lab's intricate isolating machinery.

My two mystery specimens continued to baffle me after I had them properly mounted and could view them under greater magnification with a binocular dissecting microscope. I tried to fit them into every couplet of Edwards' key to Ethiopian species of subgenus *Banksinella,* to which *Aedes circumluteolus* belongs, thinking that perhaps a more northern species had filtered down through the "tropical corridor" from equatorial regions and had just reached Ndumu. Some of those exotic species had unstriped thoraces, just like mine, but not one had wings with such gaudy veins. Though I hardly dared to believe it, I felt I must have discovered a new form of life.

Yet that would be a most unlikely possibility. This region had already been worked over thoroughly by my three expert colleagues, Botha de Meillon, Jim Muspratt, and Hugh Paterson. A novice following them was probably simply making a novice's typical sort of mistake. But I surely couldn't find the error. And I did bolster myself slightly by reflecting that probably all three of those entomologists put together had not looked at as many *A. circumluteolus* as had passed under my eyes.

This called for a conference with Botha in his Johannesburg laboratory. I told him directly what I thought, rather than challenging him blankly with the specimens, for I wanted him to take the defensive side at once. If I could crash that, everything would be fine.

He went through Edwards, just as I had done, but also brought out several trays of pinned *Banksinella* specimens from other parts of Africa. The collection at the South African Institute for Medical Research is excellent, most of it having been prepared by Botha himself. Among all this comparative material, perhaps he would find counterparts of the Ndumu wraiths.

He failed, and I waited expectantly to hear his corroborative words. "Very impressive," he said, "but not sufficiently so. Two specimens are very slim bases on which to describe a new species. If they really are something different, you'll be getting more of them. Just hold on to these and see if anything else happens."

"But I have heard of some people who have described new species on the basis of single specimens," I objected, trying not to sound plaintive. "Aren't two twice as good as that?"

"That's true enough," Botha agreed. "However, as you know, many described species eventually fall when it is discovered that they were only odd variants of something already known. That is especially common when the original series was too small. You can go ahead and describe these now if you wish—I certainly won't try to stop you. But my advice remains that you should be patient. If you get more, you'll be on safer ground."

"How many would you recommend?" I asked.

"Oh, I don't know exactly how many," he replied. "Just wait and see what happens, as I said before."

I could hardly endure the interval until our next scheduled expedition to Ndumu. At this "session" I must surely find additional specimens, for I reasoned that I had probably overlooked many of them before chancing to notice their slight but peculiar differences. It is common experience that we see best that which we are trained to see and for the rest are sometimes as good as blind.

One might want to know why I was so keen to describe a new species. That was no mystery at all to me. In my teens I had made my first contact with professional biologists at The Academy of Natural Sciences of Philadelphia. Although I did not realize it then, this was just before the end of a long era during which the chief function of natural history museums was to announce the discovery of new kinds of animals and plants to the world. Thus the people I met—and immediately worshiped—were practically all of the "old fossil" school, some of them with hundreds of discoveries to their credit. When these scientists had been relatively young men, each had chosen a certain fairly restricted field in which to specialize so that he could entertain reasonable hopes of becoming an authority in it on a hemispherical or even global basis. There was Henry W. Fowler with his fishes, Ezra Cresson with his wasps, Francis Pennell with his figworts, Henry A. Pilsbry with his mollusks, and Morgan Hebard with his grasshoppers. Who, in his young right mind, would not want to be like those masters?

Several concurrently developing factors combined to push that kind of natural history into the background. Mind you, it is still going on, as it surely must, but it is now far from the dominating urge in the progress of a good natural history museum. One of the most obvious causes of the decline in emphasis on taxonomy was the fact that several centuries of concentrated work had led to the discovery of almost all conspicuous animals and plants. Probably a million or more forms of life remain to be named, but they become, with each passing decade, progressively smaller, less important to mankind, more difficult to locate, and exasperating technically to study. The only reason so much is known about the microscopic characteristics of the sexual appendages of male mosquitoes is that we are vitally interested in that group of gnats.

But the world was now moving along faster than it ever had before and biology suddenly flowered into a living science. Ecology and ethology were born. People wanted to know how organisms managed to get along in their environments and how they behaved. When persons with enough curiosity went to natural history museums to find out, all they saw were glass-fronted cases containing musty, stuffed specimens, and there was nobody visible to tell them anything. The authorities were in that sacred precinct called "behind the scenes," and if one did manage to make his way past sentries to any of those cubicles, the incumbent savant was likely to know very little, if anything, about the habits and associations of his special subjects. "Only thing I can tell you about this specimen," he might say, "is that it arrived from Sumatra in a bottle."

Museums began to go through a thorough metamorphosis, both externally and internally. The old stuffed specimens in glass cases were replaced by the "habitat group," which displayed an instant in the lives of animals and plants at a definite time and place. As "old fossils" died, they were likely to be replaced by young Ph.D.'s with training in animate aspects of biology. No longer did the specialists remain behind the scenes. Instead, one could often find them lecturing to lay audiences or in conference with representatives from industrial and pharmaceutical firms or government agencies. These changes had been hastened by a redistribution of money. Patrons rich enough to fully support private institutions had become scarce. Museums now had to improve their public images in order to command admission fees. And by giving a more practical direction to their research (though it toppled many an ivory tower), they drew subsidies from big business.

I recognized all this as good, wise, sound—and inevitable. But I regretted it and still do in a purely personal and secret way. To me it seemed as if the old museums had been destroyed. Surely the pure academic atmosphere would never prevail there again, for applied science must be served, too. While it does no good to weep for the dead, one nevertheless indulges in it. And so, with all my good sense telling me that it was utter nonsense, I desperately wanted to discover and describe a new species of some kind on my very own, in the same spirit that my early heroes had done so.

A good dash of vanity was involved as well. To illustrate: if I were to refer to *C. theileri* formally and in full, I should have to write the name as *"Culex (Culex) theileri* Theobald." Where does Theobald come into it? He, of course, is the entomologist who published the first recognizable description of that species, and his name will be linked with *C. theileri* forever. It is rather ironic that a man can do much greater things in our society and be quickly forgotten although his work remains permanently with us. Perhaps we ought to adopt rules of language or convention such that one always says, "I attacked the can of beans with a can opener Smith," or "I fastened the baby's diaper with a safety pin Jones." Anyhow, you can see where my two little mosquitoes might lead, provided I could muster a few others like them.

Theobald, by the way, was very famous indeed. I made a tabulation of all the species of mosquito that ABVRU's various workers had encountered in Natal province and found that among a total of eighty-nine, Theobald's name followed forty-two. His successor at the British Museum (Natural History) was F. W. Edwards, whose book was my mainstay, and that gentleman did very well, too, with twenty-seven species bearing his stamp. The remaining twenty were credited to almost as many people, though Botha was associated with three of them and Jim Muspratt had described two. Even good, venerable Linnaeus was represented by *Culex pipiens,* the common household gnat. Obviously I was hoping to invite myself into a sparkling company.

Mosquito-sorting took much longer henceforth. I lingered over each specimen of *A. circumluteolus,* and I drove the boys mad with my insistence that the tubes be absolutely gleaming. On the first return session at Ndumu, I rapidly fell into despondency, for my hypothesis of having previously "overlooked" the supposed new species fell flat. Each *A. circumluteolus* was exactly as it should have been, with boldly black and yellow-striped throax and conservatively marked wing-veins. During that session, and in others to follow, *A. circumluteolus* passed through my hands one by one, building up to their thousands upon thousands as I approached my eventual total of 81,702, but, to use Botha's expression, "nothing happened."

I had almost resigned myself to his inference that my two pinned novelties might not really be "something different," but only strange variations from normal *A. circumluteolus* stock. However, if that were true, I should be finding intermediate examples. If dogs and cats were all of one species, the two forms that we know by those names representing only the extremes of possible variation within the genetic pool, we would constantly encounter all degrees of doglike cats and catlike dogs, at the median point finding a dog–cat that we could not ascribe to either side. But *A. circumluteolus* was either pure cat–cat or pristine dog–dog, whichever you prefer: no genetic spillover showed up anywhere along the line.

Then in November, ten long months after the original discovery, one of my mosquito-catchers, Dom Dom, brought in a box of tubes that contained two additional specimens of the marvelous creature. One was moderately worn, but even that was advantageous in a way, for despite such an esthetic drawback it was still readily identifiable as an obvious sister belonging to what had now become a quartet that stood apart from *A. circumluteolus* with all the dignity of self-determination.

"You can't miss," said Botha after he had studied the new pair and compared them with the originals. "Of course they are all females. It would be preferable to wait until you get some males and, better yet, larvae as well. But that might not happen in the limited time you will still be here, and these four, conforming to each other as perfectly as they do, should be enough to convince anybody. At least *I* am convinced. You might as well go ahead and publish."

Did I "go ahead"! I wrote the paper that day, and it appeared not long thereafter in *The Journal of the Entomological Society of Southern Africa.* Of course I had to find a name for the new species, but that was easy. Golden-scaled wing-veins were its outstanding characteristic, so it must be called *"aurovenatus."* As taxonomists do so often, they had recently discarded a subgeneric designation; *Banksinella* was replaced by *Neomelaniconion,* which was belatedly found to have priority in published entomological literature. Thus I came to occupy an immortal niche with the following inscription: *Aedes (Neomelaniconion) aurovenatus* Worth. Greetings, Theobald! Greetings, my deceased mentors at the Philadelphia Academy!

Of course it is not quite as cut-and-dried as that. Publication serves only to announce that a scientist believes he has discovered a new species. If no one accepts the discovery, and particularly if the most renowned experts actively repudiate it in print, the bright little bubble soon bursts and no trace remains except for a couple of wasted pages of type in an old journal.

But how were the experts to judge? They wouldn't all converge on Johannesburg just to peer at four tiny mosquitoes that I had been arrogant enough to endow with a golden name. Botha discussed the problem with me from several standpoints. One must distribute these specimens, not only so that they would be accessible for study by other scholars, but also as a matter of prudence. For example, during World War II several important museums in Middle Europe had been destroyed, and irreplaceable "type" materials were lost. "Types" are the specimens on which a scientist bases his original description of a species. He actually selects a single individual and calls it the type, while if he has a series of individuals, the remaining ones are called "paratypes." All of these then become permanent standards, like that international yardstick the platinum meter in Paris.

"I suggest that you leave the type and one paratype here in the collection of the South African Institute for Medical Research," said Botha. "But you better deposit the others somewhere else. This place could burn down, you know. Why don't you send one to Peter Mattingly at the British Museum [Natural History] and the other to Alan Stone at your U.S. National Museum?"

"Do you think they would really want them?" I asked.

"Of course they would," said Botha. "They'll be very glad indeed."

Mattingly wrote me a very kind letter of acknowledgment, as if he truly welcomed the arrival of that small package. "Generosity" was one of the words he used. As for Alan Stone, he was an old friend, and I had decided to wait until my return to the States, when I would give the paratype to him in person. *That* small package traveled in my suitcase in a long roundabout way as I made the journey home from South Africa via India, Malaya, Japan, Alaska, and California. Not many mosquitoes have flown that far.

As soon as I could, I hopped a train from Philadelphia to Washington and made my way to Alan's office in a wing of the National Museum. Alas! He was on vacation. I carefully undid the wrappings of the box and peeped inside to be sure that the mosquito was still in one piece after all its peregrinations. Finding that everything was in order, I entrusted the rewrapped parcel to an assistant, who promised to bring it to Dr. Stone's attention.

Several weeks later I received his confirmation of receipt, but my dreams of immortality were suddenly shocked into that aftermath of most dreams: disillusionment. Alan had studied the specimen but could not see how it differed from *Aedes luteolateralis*—another species in Edwards' key, though I had checked that one and decided against it. What, asked Alan, had led me to think that the specimen was unique? Could I send him a reprint of my description?

My reprints were all boxed up and on shipboard, being transferred from South Africa to my next station in Trinidad. As I sat and read Alan's letter, all I could rely upon was memory. What would be the most telling point?

Finally I settled on Vein 3, a short vein near the wing-tip. This one was dark-scaled in all other Neomelaniconions but stood out particularly brilliantly in *A. aurovenatus*, with a thick double row of golden scales along its entire extent. I posted the letter and then waited. This would be the ultimate test, for if Alan remained unpersuaded, the rest of the world would quickly accept his denunciation.

But all was well. "I overlooked that vein," he wrote.

COMING OF THE GIANT WASPS, BY LOREN EISELEY

When the great solitary hunters known as cicada-killers came to his backyard, provisioning their nests with prey as large as themselves, the observer found that their mysterious ways defy the simplistic human view of evolutionary events.

I will never forget that autumn, just because of the warm September light on the hillside, and the way those great sphex wasps, the cicada-killers, came in over the grass like homing bombers. Perhaps they had hunted briefly about in other summers, but this year someone had told me moles were tunneling in the rear slope of my apartment house. I had gone out to investigate and had needed only one glance at a quaint, big-lensed, intelligent head coming out of one of the burrows to know that something magnificent had intruded into our little backyard garden.

As the days passed one could hear an occasional cicada's song terminating abruptly in a kind of stifled shriek—a sign that an assassin had reached him. If I waited quietly it would not be long before one of the huge wasps, the largest on the eastern seaboard, would come gliding in toward its excavation with a paralyzed cicada in its clutches. Then the creature would emerge alone, circle at speed to get its bearings, and wing off into the trees.

What I admired about the wasps was the deadly perfection of instinct they exhibited, and they utter indifference to man. If I stood near one of the holes long enough, I might be circled but never attacked. I was merely being utilized as a beacon marker. Unfortunately, for that purpose I was far too shifting and impermanent.

These wasps of the genus *Sphecius* are solitary, and giants of their kind. The first I ever saw, years ago in another state, was carrying prey heavier than herself and getting it airborne by utilizing the corner of a building to bounce her upward-thrusting legs against, while using her wings at full power. When she reached sufficient altitude she zoomed off in a beeline for her distant burrow. Here, in the backyard, the wasps had selected the sunny hill for their nests. This neighborhood activity is their only sign of incipient socialization. They had come to this spot, I suspected, because there was little or no appropriate wildland remaining in the growing suburbs around us. If they took cognizance of human beings at all their would not have intruded here. They frightened old ladies by their mere size, and gardeners trampled remorselessly upon their excavations. But still, in the early morning light, they would return to dig, and their energy was boundless.

I was as pleased that autumn as though someone had reported a panther in our pine tree. I was getting old enough to want to rethink what I had learned when I was younger. I believe now that it was the coming of the giant wasps that first implanted some doubts in my mind about the naturalness of nature, or at least nature as she may be interpreted in the laboratory.

These wasps and their assorted brethren, the tarantula-killers, present in miniature several of the greatest problems in the universe. Beautiful as they appear in sunlight, their deeds below ground are less edifying. They would justify Darwin's well-known remark about the horribly cruel works of nature, or even Emerson's observation that there is a crack in everything God has made.

Yet these hymenopterans exercise no moral choice. Their larval stage has to be supported by a peculiarly frightful form of feeding upon paralyzed flesh, while the adults must contribute to their survival by one of the most precise surgical operations known among the lower creatures. Otherwise the species would perish. Beauty and evil, at least by human standards, course together over the autumn grass. The wasps have always impressed me as formations of fast fighter planes impress me—incredible beauty linked to destruction in the heart of man. As a cynical pilot once put it of his charges, "They are nothing but a flying gun platform."

Similarly one could say of *Sphecius speciosus* that it is nothing but a flying hypodermic needle. But how wrong, how terribly wrong, oversimplified, and reductionist would be both observations. Into man's jet fighters, year by year, have gone some of the most elaborate scientific calculations in the world—mathematical equations of the utmost abstract beauty which, when embodied in reality, result in speeds outrunning the human imagination. Their potential uses of terror lie in the ambivalent nature of man, or perhaps in the ambivalent nature that created man. So the great wasps, the invaders of the autumn grass, carry navigational aids whose complexity remains unexplained and whose surgical intent is comprehended, if at all, only in the dream that lies below all living nature—a dream as tenuous and insubstantial as the shaft of September light through which roams a flying deadly lancet. "My thoughts are not your thoughts," runs the Biblical injunction of Jehovah. "Your ways are not my ways. I make the good. I create evil. I, the Lord, do all these things."

I have come to believe that in the world there is nothing to explain the world, nothing in nature that can separate the existent from the potential. I start with that. Biological scientists, however, are involved by necessity in the explanation of life. In the end many are forced into metaphysical positions which reflect their own temperamental bent. There are reductionists like Jacques Loeb, who strove to bring life into the management compass of physics and chemistry, or men such as the philosopher Henri Bergson, who attempted to distinguish life as a separate, indefinable principle, the *élan vital*. Between these extremes we all flounder, choosing to close our eyes to ultimate questions and proceeding, instead, with classification and experiment. Even then our experiments are apt to be colored by what we subconsciously believe or hope.

This conflict extended into the last century, intensifying with the discovery of the principles of evolution. One of the last great evolutionary controversies of that century arose, in fact, over the behavior of the solitary wasps whose mysterious habits, so Jean Henri Fabre, the French entomologist, proclaimed, simply did not lend themselves to an explanation by means of the selection of chance Darwinian mutations. To my mind the controversy was never fully resolved, only softened and eventually dropped as other more comprehensible discoveries diverted the naturalists of the new century. Nevertheless, the world owes a debt to Fabre, the impoverished scientist who worked all his life in the sandy stretches of southern France. He was too unschooled to accept readily what he was told in other people's books, including Darwin's. Instead he lay under the spell of the elegant French experimentalists who preferred controlled investigation to armchair theorizing.

It is not sufficient to say, therefore, that the schoolmaster Fabre was an anti-Darwinian who saw, in the perfection of instinct, an utter barrier to evolution. Fabre merely chose, on the basis of his field studies, to ask some legitimate and penetrating questions. Neither can his American successors in wasp studies, George and Elizabeth Peckham, be regarded as having totally overthrown the views of the French entomologist. Actually there is an overlap of complementary observation which does not justify the total Darwinian triumph implied at the end of the century by Darwin's devoted follower Edward Poulton.

Darwin himself realized that among the amazing life cycles of insects there were cases difficult to explain on the basis of pure undirected variation. At one point he confessed in *The Origin of Species* that many instincts could be opposed to the theory of natural selection, "cases in which we cannnot see how an instinct could possibly have originated" and "in which no intermediate gradations are known to exist." It was this sort of problem, notably the knowledge possessed by the sphex wasps of their opponents' neurological weaknesses, which had troubled Fabre.

Poulton contended that the Peckhams' discovery of instances of variable behavior among wasps revealed that instinct had its imperfections and that complicated behavior could still be accounted for in the Darwinian empire of accident and natural selection. Poulton was not as fair as Darwin. He confined himself to general references to the Peckhams' work and did not try to answer Fabre's specific questions.

A careful reading of Fabre reveals that he was not unaware of variation in the behavior of his wasps. He was, nevertheless, notably impressed by the surgical knowledge manifested by both grub and adult, instincts which seemed to rule out any theory of their origin through the selection of chance variations. The other Darwinian refuge of that time lay in the suggestion that learned behavior, "habit," might precede and prepare the way for the emergence of purely instinctive behavior.

The Peckhams, in their studies of the American solitary wasps, were concentrating heavily upon the variability which by then was the preoccupation of every naturalist. Fabre, by contrast, had expressed doubts, not totally answered to this day, as to how unaided Darwinian natural selection or, indeed, selected "habit" could produce something that would be of no natural use until the chain of instinctive reflexes led to the survival of the wasp. Such survival could only be effected through a very complicated mosaic of perfected and interlocking behavior distributed between the adult insect and its larval offspring. To paraphrase one modern naturalist, John Crompton, a surgeon does not learn his trade by indiscriminately pursuing and slashing at his potential clients with a sharpened lancet. Neither is it likely that the sphex wasps acquired their skill through chance behavior which, in the most successful, slowly froze into the rigidities of perfected instinct.

The Peckhams reported triumphantly that they had observed potter wasps who sealed cells without including eggs or the usual paralyzed spiders within them. Thus they implied variation and instances of failure—the test from which emerges selective evolution. But there is no account as to whether those particular females were gravid. It may well be that they were merely giving vent to exuberant production of their mud cells just as, on a higher level, wrens have a habit of constructing nests they often never utilize. This overplay, a kind of psychological release, is entwined with many forms of instinct, as when a shepherd dog tries to herd people or a flock of daisies—instances which have been reliably observed.

The fearsome operations of the sphex wasps depend upon an uncanny knowledge of the location of the nerve centers of their prey in order to stun, not kill the creature. The larvae, also, must possess an instinctive knowledge of how to eat, in order to prolong the life of the paralyzed body which they devour. To complicate matters further, the victim, even formidable, outsize victims like tarantulas, seems to have some innate foreknowledge of its helplessness, some fear of which its agile opponent takes absolute confident advantage.

All is arranged in such a manner as to suggest the victim possesses an innate awareness of his role, but cannot evade it. If, so Fabre muses, pure chance has, through long ages, decreed this relationship between hunter and hunted, why have not the cicada, the cricket, and the tarantula equally evolved a defense against their fate? If we attribute success to natural selection in the case of the wasp, why has not the same power been at work for the victim? Or, on the stage of life does the victim play a foreordained role, as well as the huntress?

Whatever forces have been at work in the evolution of the wasp family, it is clear that they have little, if anything, to do with that nineteenth-century cliché about "the effects of habit," which tells us nothing. Fabre was right in that judgment, possibly right even in his fateful admonition, "It is not in chance that we will find the key to such harmonies. The man grappling with reality," he concludes, "fails to find a serious explanation of anything whatsoever that he sees."

Let us grant that Fabre chose not to explore the evolutionary road. Let us admit that his metaphysical bent lay in another direction. But the attempt of many of the Darwinian circle to explain the mysteries of instinct was not always enlightening. They confused their own Darwinism by choosing the best of both worlds when they argued that chance-acquired habits might sink into the germ plasm. The experimenter on his little patch of poverty-ridden soil at Sérignan had toiled long enough to know that the world he investigated provided more mysteries than answers. In his old age he adhered to that conclusion. Perhaps it was his philosophical weakness. Perhaps, on the other hand, it was simple honesty.

The inorganic world out of which life has emerged and into which, in season, it falls back, possesses the latent capacity for endless ramification and diversity. A few chance elements which appear thoroughly stable in their reactions dress up as for a masked ball and go strolling, predator and prey together. Their forms alter through the ages. They are shape-shifters, role-changers. Like flying lizards or ancestral men, they run their course and vanish, never to return. The chemicals of which their bodies were composed lie all about us, but by no known magic can we return a lost species to life. Life, in fact, is the product of singular and unreturning contingencies of which the inorganic world disclaims knowledge. Only its elements, swept up in that mysterious living vortex, evoke new forms, new habits, and new thoughts.

I am an evolutionist. I believe my great backyard sphexes have evolved like other creatures. But watching them in the October light as one circles my head in curiosity, I can only repeat my dictum softly: In the world there is nothing to explain the world. Nothing to explain the necessity of life, nothing to explain the hunger of the elements to become life, nothing to explain why the stolid realm of rock and soil and mineral should diversify itself into beauty, terror, and uncertainty. To bring organic novelty into existence, to create pain, injustice, joy, demands more than we can discern in the nature that we analyze so completely. Worship, then, like the Maya, the unknown zero, the procession of the time-bearing gods. The equation that can explain why a mere sphex wasp contains in her minute head the ganglionic center of her prey has still to be written. There is nothing below a certain depth that is truly explanatory. It is as if matter dreamed and muttered in its sleep. But why and for what reason it dreams, there is no evidence.

It is now high autumn. Apples are falling untended and smashing on the stones I have come to call Wasp Alley. The smell is drunken, ciderous. In the growing dark wasps of many species—vespas, yellowjackets, mud-daubers— clamber over the ripe ungathered fruit. On this particular evening something more formidable rises and bumps my nose inquisitively before it flies away over the roofs. It is one of the giant sphexes caught in an innocent moment of adult feeding, the deadly needle sheathed at last. Instinctively I know this will be our final encounter of the haunted year.

But still, not quite. The sun, a week later, falls in gold October splendor over the little hillside. Coming home in the afternoon I sit down, a little stiffly, and survey the drowsy slope where the closed burrows of the sphecoids are hidden in the autumn grass. At the bottom of each burrow reposes a mummy case, a sleeping pupa. It will lie there still drowsing under the winter snows, and surrounded by the emptied husks of its feeding.

Beneath the late spring sunlight of another year a molecular alarm will sound in the coffin at rest in that silent chamber; the sarcophagus will split. In the depths of the tomb a great yellow-and-black sphex will appear. The clock in its body will tell it to hasten, hasten up the passage to the surface.

On that brief journey the wasp may well trip over the body of its own true mother—if this was her last burrow—a tomb for life and a tomb for death. Here the generations do not recognize each other; it remains only to tear open the doorway and rush upward into the sun. The dead past, its husks, its withered wings, are cast aside, scrambled over, in the frantic moment of resurrection.

The tomb has burst. A tiny chain of genes and releaser genes in the black dark has informed the great winged creature of her destiny, the unseen flowers, the shrilling of cicadas in the sun. She carries not alone the surgical instrument, but the map of operations yet to be performed on an insect she has never seen. She is a nectar feeder, but it is for carnivorous grubs that she will labor, the grubs of which she was once one, feeding on paralyzed flesh in the sightless gloom of a walled chamber. Here beneath the leaves on the autumn grass slept nature, or a part of nature so beautifully, so exquisitely contrived that it was hard to imagine error, hard to conceive of all the pieces of that intricate puzzle being put together from the blind play of natural selection alone. Looked at from one point of view, nature had created monstrous evil, the tormenting of helpless, paralyzed flesh. Looked at in another way, the eternal storm maintained its balance.

I remembered how that formidable autumn creature had hovered before my face as though questioning my own existence in the apple-strewn twilight. Apples were still falling untended while far away, on another part of the planet, people died of hunger. The great sphex itself was doomed in the oncoming frosts of autumn. Everything living was falling, disintegrating as under the violence of an unseen hurricane.

"Created to no purpose by an endlessly revised genetic alphabet," one part of my mind contended. "A work of ecstasy," the words of Emerson echoed in another chamber of my thought. Throughout September I had watched the tiger-faced sphexes digging with furious energy. I had heard the muffled shriek that ended the cicada's song. On that lonely backyard slope it had somehow pleased me that the wasps came and went as though I belonged to another world they chose to ignore, a misty world for which they carried no instruction, just as I carried none for the totality of the night. Though shorn of knowledge, willing to accept the dreadful otherness of the Biblical challenge, "Your ways are not my ways," I had come to feel at last that the human version of evolutionary events was perhaps too simplistic for belief.

There is a persistent adage in science that one must not multiply hypotheses unduly and without reason. I grant its usefulness. Nevertheless, it can sometimes lead to the assumption that science finds nature simple and that someday all will be known. Vain delusion, incredible folly, I thought, brooding there at sundown over the sleeping surgeons known as sphex. We, our species, will be gone before we know.

I drew my aching foot beneath me and tried once more, in human terms, to balance the unfathomable world. As in the case of Henri Fabre on his little sand plot, uncounted mysteries had away of persistently intruding into my mind. The wasps' master chart of surgery was not always perfect. Still, it was terrifying enough to provoke the envy of any practicing physician. This evolutionary marvel was not just that of slow selection for size or greater running speed, as among horses. The entire pattern had to work or the species would perish. There seemed to be no intermediate possibility. The larvae have to feed in a certain way. The adult female has to seek prey upon which it has never personally fed. It has, furthermore, to identify that prey. The wasp has to bring its paralyzed cicada back over a distance to a burrow it has already constructed and whose position it has previously mapped with the care of an aerial navigator. To explain this uncanny phenomenon by computerized armchair genetics may be theoretically possible if one starts from certain current assumptions which leave me vaguely uncomfortable. Perhaps that can be termed my metaphysical position. I am simply baffled. I know these creatures have been shaped in the cellars of time. It is the method that troubles me.

Some ten years after Fabre's death in 1915, Alexander Petrunkevitch, the spider specialist, had described his own adventures with a tarantula-killing wasp, *Pepsis marginata.* All of the great wasps are fascinating in their diverse surgical habits, but what had long intrigued me about this particular account of the Caribbean killer wasp, *Pepsis,* was something that seemed once more to lie doubly out of time and belief. Shifting my foot again in the misty light on the hillside, I tried to recall it. It was important now; I had not many more autumns in which to ponder such problems. *Pepsis*, the tarantula-killer, dueled with a far more formidable creature than a cicada. Its knowledge of tarantula anatomy was just as deadly as that of the giant sphex. But one thing more, one bit of preternatural intelligence continued to challenge my faith in the pure undiluted chances of natural selection.

When *Pepsis* paralyzed her giant foe and deposited her egg, she added one more complicated pattern to the behavior of the killer wasps. She packed her big, hairy opponent so masterfully into its grave that it could never dig its way out even if it were, by some chance, to recover. Every limb of the huge spider was literally handcuffed to earth. Poison needle, utter paralysis, were not enough. A final act of devastating ingenuity had been added.

The autumn light was growing dull about me, the shadows were gathering. I was beyond the country of common belief; that would seem to be the source of my problem. I had spent a lifetime exploring questions for which I no longer pretended to have answers, or to fully accept the answers of others. I was slowly growing as insubstantial as the sunlight on the hillside. I could not account for myself any more than I could validate in material terms the strange anatomical charts that slept, for now inactivated, in the tombs beneath my feet.

Slowly, painfully, I arose and limped away. As I walked I knew, with the chill of a not-too-welcome discovery, that I was leaving the sharply defined country of youth and scientific certitude. I was seeking an undiscoverable place, glimpsed long ago by the poet Shelley,

> built beyond mortal thought
> far in the unapparent.

Strangely, in a little-known passage, the great experimentalist Claude Bernard once echoed, more grimly, the same idea. "I put up with ignorance," he said. "That is my philosophy." Thus ended the visitation of the giant wasps. I never saw them again.

THE AUTHORS

ROBERT BELOUS has one of the most enviable jobs in the vast federal bureaucracy. Originally from New York City, and a nuclear engineer by training, he decided several years ago to turn his lifelong avocation of nature photography into a full-time career. This led to a post with the National Park Service, and specifically with the Alaska Task Force, which is responsible for planning eleven proposed new park preserves in the forty-ninth state. This means that, as chief of the photographic unit at the Anchorage field office, Belous has been exploring with his cameras some of the most wildly beautiful land in North America. Besides *Audubon,* his pictures and stories have appeared in *National Geographic, National Wildlife, Field & Stream,* and many other publications.

WADE T. BLEDSOE, JR., has spent the past several summers living amidst the largest congregation of the largest land-based carnivore on Earth—the brown bear of Alaska. As a doctoral candidate in wildlife ecology at Utah State University, he and his colleagues have flown each June to the McNeil River, an undistinguished stream on the Alaska Peninsula that flows into Cook Inlet two hundred miles southwest of Anchorage. The McNeil is undistinguished, that is, until upward of eighty brown bears—adults, yearlings, and new cubs—gather at its falls for a month or more to feast on spawning salmon. Normally solitary giants, the bears are suddenly thrust into a social situation that produces fascinating behavior ranging from violent battles to cub swapping. A Texas native, Bledsoe earned a master's degree in zoology from East Texas State University for a behavioral study of captive baboons.

HAL BORLAND, novelist, essayist, and naturalist, was born in Nebraska, grew up on the Colorado plains, the son of a small-town newspaper editor, and has lived for the past thirty years on an old farm in the Berkshire Hills of northwest Connecticut. The story of his boyhood is told in two autobiographical books, *High, Wide and Lonesome* and *Country Editor's Boy*. His novels also are laid in the West, most notably *When the Legends Die*, considered a modern classic. And he has written of the natural history of New England in many books, including *Hill Country Harvest*, which was awarded the John Burroughs Medal. Each Sunday Borland contributes a "nature editorial" to *The New York Times*; he is an *Audubon* contributing editor; and he writes a quarterly essay for *The Progressive*, as well as a weekly column for several New England newspapers. His most recent work is *Hal Borland's Book of Days*.

ARCHIE CARR is eminently qualified to write in defense of snakes, for not only is he a leading expert on the subject of reptiles, but he is one of those rare scientists who possess great literary gifts. Attesting to the latter are such honors as the John Burroughs Medal for exemplary nature writing and an O. Henry Prize for one of his short stories. A boyhood fascination with turtles and snakes, which were abundant in those days in his native Florida, led him to study zoology at the University of Florida, and he has taught on the Gainesville campus ever since he received his Ph.D. in 1937. He is now graduate research professor as well as curator of zoology for the Florida State Museum. But Dr. Carr's renown carries far beyond his home state, for he is the world's foremost authority on sea turtles, the subject of his book *So Excellent a Fishe*. He is executive director of the Caribbean Conservation Foundation, an organization founded to save the green turtle from extinction, and he has led expeditions to several continents to study these great ocean-going reptiles.

LOREN EISELEY was exposed early in his boyhood, on the Great Plains, to the magic of poetry and the beauty of nature, for his father was a one-time itinerant actor and his mother an artist. In the difficult years of the Depression the course of his life wandered from drifter to fossil hunter to sometime college student, and ultimately to a distinguished career in science. The University of Pennsylvania honored him with a special chair as Benjamin Franklin Professor of Anthropology and the History of Science. His poetic writings, including *The Immense Journey* and *The Invisible Pyramid*, have earned him many honors, most notably election as a member of the National Institute of Arts and Letters. Dr. Eiseley's most recent book is his autobiography, *All the Strange Hours*.

COREY FORD really couldn't be considered a nature writer in the strict meaning of the term. The story reprinted in this collection is probably the only one he ever wrote for a natural-history magazine, but it was adapted from a book that will long stand as one of the truly great classics of its kind: *Where the Sea Breaks Its Back*, the epic of pioneer naturalist Georg Wilhelm Steller and the discovery of Alaska. Moreover, Corey Ford, when he died in 1969 at the age of sixty-seven, left behind whole generations of fans who appreciate the out-of-doors—the readers of *Field & Stream,* where, since 1953, he had written each month about the misadventures of the fictional Lower Forty Shooting, Angling and Inside Straight Club. A fine and often funny writer Corey Ford was, too, as readers of *The New Yorker, The Saturday Evening Post,* and many other publications knew, for since the late 1920s he had produced more than five hundred stories and thirty books on subjects ranging from social commentary to the history of American intelligence operations beginning with the Revolution. *The Best of Corey Ford,* a collection of his life's work, edited by Jack Samson, was recently published. One essay, his last, called "The Road to Tinhamtown," shows "how fine a writer was this gentle man."

SYLVIA FOSTER, to the astonishment of those readers who admire her eloquent but all-too-infrequent contributions to *Audubon,* is not a professional writer, for the story included herein is only her second to be published. One reason, which is immediately obvious, is that she spends months—sometimes years—becoming acquainted with her subjects before putting words to paper. A former housewife, and then a waitress, she is now attending college full-time and spending all her free hours in what she calls "deserting": hiking, backpacking, exploring, observing in the wilderness of the Sonoran Desert near her Tucson, Arizona, home.

C. E. GILLHAM had "been around" the wildlife business in the seventy-two years prior to his death in 1970. He was a hired bounty hunter who tracked down and killed the only white wolf ever seen in Arizona. He was the government waterfowl biologist for the Mississippi Flyway, and made ornithological history when he discovered the nesting grounds of Ross's goose in the Canadian Arctic. And he was the Alaska territorial biologist in the years before statehood. A prolific author, Charley Gillham wrote several books on the Far North and perhaps two hundred stories for out-of-doors magazines. But in his later years he found he could no longer shoot wildlife for sport—or for spite, as the tale included in this collection proves.

LOUIS J. HALLE has had a many-faceted and most fascinating career. Among naturalists, he is noted for his evocative and insightful writings. Perhaps his best-known nature book is *Spring in Washington,* a classic that is still widely read thirty years after its publication. His most recent work is *The Sea and The Ice: A Naturalist in Antarctica,* which was adapted for a special issue of *Audubon.* But Halle's primary career has been in the delicate arena of international relations. He served on the policy planning staff under Secretaries of State Dean Acheson and John Foster Dulles; and from 1956 until his retirement two years ago, he was director of strategic studies (the problems of war and peace) at the Graduate Institute of International Studies in Geneva, Switzerland. His many writings in this field include another classic book, *The Cold War as History.*

JOSEPH WOOD KRUTCH followed many distinguished and varied careers during his seventy-six years—as teacher, drama critic, philosopher, man of letters—and as one of America's most articulate spokesmen for nature and conservation.

From 1924 to 1950 he was drama critic for *The Nation,* and from 1943 professor of dramatic literature at Columbia University. He wrote books on literature and drama, on Samuel Johnson, Edgar Allan Poe, and Henry David Thoreau, and then in 1950 his life took a new direction. Retiring from his teaching post and from *The Nation,* he moved to an adobe brick home in Arizona and wrote *The Desert Year,* the first of several books on the Southwest wilderness. Among his many honors in his later years was the Emerson-Thoreau Medal of the American Academy of Arts and Sciences. Shortly before his death in 1970, Dr. Krutch wrote these last words for an Arizona newspaper: "The seventies may be the beginning of the end, or the beginning of a new civilization. If it becomes the latter, it will not be because we have walked on the moon and learned how to tinker with the genes of unborn children, but because we have come to realize that wealth, power, and even knowledge are not good in themselves but only the instruments of good or evil."

GEORGE LAYCOCK is an *Audubon* field editor—one of a small, select team of highly respected writers who are regularly assigned by the magazine's editors to report on important wildlife and environmental issues. Laycock, who has a degree in wildlife management from Ohio State University, has traveled for *Audubon* from the northern coast of Alaska to the Caribbean to the Leeward Islands of Hawaii in search of stories involving, most often, America's wildlife heritage—the pesticide poisoning of pelicans, the conflict between feral burros and rare desert bighorn sheep, illegal traffic in alligator hides, the poaching of waterfowl, the testing of nuclear bombs in a wildlife refuge, and, among many others, the continuing conflict between man and coyote. He has more than thirty books for both adult and young readers to his credit, including, most recently, *Autumn of the Eagle,* which was nominated for a National Book Award, and *Bird Watcher's Bible.*

JOHN MADSON, if his hectic schedule allowed him to spend more uninterrupted time at the typewriter, would surely be one of America's best-known nature and conservation writers. As it is, Madson fans—and there are many—must be content with his reasonably regular contributions to the pages of *Audubon,* and his sharp commentaries on wildlife matters sent to local newspapers and outdoor journals in his role as assistant director of conservation for Winchester-Western, the sporting-arms company. Only one book presently bears Madson's byline: *Stories From Under the Sky,* a rarity published fifteen years ago by the Iowa State University Press. But a Madson book on the tallgrass prairie is near completion, and the subject is a natural one. For Madson calls the prairie his home; his great-grandfather settled on the Iowa frontier in 1853, and more often than not his *Audubon* articles tell about prairie country and prairie wildlife.

PETER MATTHIESSEN, of all living naturalist writers, truly merits the title of explorer. In quest of material for his evocative and poetic books and articles, including several for *Audubon,* he has dived in search of the terrifying great white shark off the coasts of Australia and Africa; explored the headwaters of the Amazon and rafted down "impassable" Andean rivers; sailed on a turtle boat in the

Caribbean; gone on expeditions to the Himalayas, New Guinea, Africa, the Arctic, and the wild Hawaiian island of Maui. To quote the publication *Current Biography:* "Because of his camera-like eye for landscape and animal life, his moral awareness, and his ability to transform experience with his own peculiar vision, Matthiessen has been compared by some critics to W. H. Hudson and Joseph Conrad." He is also an avowed conservationist, and his writings register "the cataclysmic impact of technology on natural resources and primitive people." A number of Matthiessen's natural-history books are considered classics: *The Cloud Forest, Wildlife in North America, The Tree Where Man Was Born, The Shorebirds of North America.* He is also an innovative novelist, and his most recent work, *Far Tortuga,* the product of nine years of writing and reflection, won critical acclaim.

RONALD ROOD was a college graduate mining gold in Alaska when Pearl Harbor was attacked. He joined the Air Force, flew P-51 Mustangs over Europe, then returned to school after the war to get a graduate degree in zoology. He was teaching college biology when a chance meeting with a magazine editor in 1953 headed him into a career writing about nature. And a prolific career it has been: more than three hundred articles and twenty books have carried the Rood byline. The titles of a few books reveal his unique view of the wild world: *How Do You Spank a Porcupine?; The Loon in My Bathtub; Who Wakes the Groundhog?; Animals Nobody Loves; May I Keep This Clam, Mother? It Followed Me Home.* His most recent book is *Good Things Are Happening;* for background, Rood traveled twenty-four thousand miles talking to hundreds of people about the unheralded things they are doing to save the environment and the creatures, man included, that must live in it. The Rood family's personal environment is a farm in Vermont's Green Mountains.

FRANKLIN RUSSELL, by every standard, has led a peripatetic life. He spent his boyhood on a farm near Christchurch, New Zealand, and studied in that country, in Australia, and in England. At various times he worked as a farmer, mechanic, truck driver, and streetcar conductor before settling into a career as a newspaperman and then an itinerant free-lance journalist. Always interested in natural history, he explored the Southern Alps of New Zealand, the Australian Outback, and camped across England, Scotland, Germany, Spain, Italy, and Egypt. When he settled for several years in Canada, he could not resist the lure of the Maritime Provinces and particularly the great seabird nesting colonies on islands in the Bay of Fundy, Gulf of St. Lawrence, and off the Newfoundland shores. His explorations of that often primitive region led to several evocative books, including *The Secret Islands* and *Argen the Gull,* both classics of modern nature writing. Among Russell's recent books is *Season on the Plain,* a dramatic narrative of how the lives of East African animals, large and small, are interdependent.

JACK SCHAEFER undoubtedly will always be best known for his writings about the American West. Among his stories is *Shane,* which became a classic motion picture starring Alan Ladd and Brandon de Wilde, and the children's tale of *Old*

Ramon. Schaefer, who lives in New Mexico, is still a devoted student of western history. But, as he writes in the introduction to his most recent book, *An American Bestiary,* he has begun a new voyage of discovery, "a pursuit of knowledge about those creatures who are my closest relatives, my fellow mammals." He explains: "Always I was writing about people, about us featherless bipeds who sum ourselves by genus and species as *Homo sapiens.* Any other creatures who crept in were merely stage furniture for the human drama. And then, as a writer, I came to a full stop. I had lost my innocence. I had become ashamed of my species and myself. I understood at last that despite whatever dodges of motive and intent I might cite, I was a contributing part of the heedless human onrush that was ruining the land I loved and forcing toward extinction ever more of my fellow creatures whose companion right to continued existence ought to be respected."

EDWIN WAY TEALE is the only naturalist ever to receive a Pulitzer Prize. That singular honor was accorded his four-volume series *The American Seasons* for which the author and his wife, Nellie, over a span of twenty years, traveled one hundred thousand miles up and down and across the United States, following and recording the passage of spring, summer, autumn, and winter. And that imaginative undertaking led to an eleven-thousand-mile journey through *Springtime in Britain,* with nostalgic pilgrimages to the countrysides of the literary-naturalist heroes of his youth such as W. H. Hudson, Charles Darwin, and Gilbert White. The author of twenty-eight books, Teale for the past seventeen years has lived on an old farm of one hundred and thirty acres in the northeast corner of Connecticut. Trail Wood, his personal sanctuary, contains wild meadows, woodlands, a pond, brooks, and a host of living things about which he wrote in his most recent book, *A Naturalist Buys an Old Farm.* This fall he will publish a five-hundred-page selection of his favorite writings from the four-season odyssey.

JOHN K. TERRES, who charted the editorial course of *Audubon* for twelve years, is the author or editor of some fifty books on natural history, including *Songbirds in Your Garden,* which has sold several hundred thousand copies. A former biologist with the U.S. Department of Agriculture and an outstanding field naturalist, Terres observed for eight years, by day and night, the wildlife on an abandoned farm owned by the University of North Carolina, in Chapel Hill. Those experiences led to his book *From Laurel Hill to Siler's Bog,* which not only earned a John Burroughs Medal but influenced the university trustees to set aside a large part of the farm as a wildlife sanctuary. Such conservation achievements by Terres are not unusual. An article he wrote for *Audubon* in 1952 saved Island Beach in New Jersey from real estate development; and he was responsible for persuading the management of the Empire State Building to turn off the tower beacon lights during spring and fall migrations to save thousands of birds from death. Terres is now completing his lifework, a one-million-word *Encyclopedia of North American Birds.*

ROY VONTOBEL spent his Connecticut boyhood roaming woodlands, marshes, and the shores of Long Island Sound—and exploring natural-history museums at every opportunity. He worked as a zookeeper, skin diver, newspaper reporter, and,

191

after graduation from Columbia University with a degree in anthropology, as a member of the editorial staffs of the magazines *Natural History* and *Nature Canada*. In the fall of 1973 he was employed by the Fisheries Research Board of Canada as a biological technician aboard the ship *Arctic Endeavor* on a forty-five-day voyage to census whale populations in the North Atlantic, Baffin Bay, and off Greenland—an expedition that led to his contribution to this anthology. Vontobel currently is working for the Archaeological Survey at the National Museum of Canada, in Ottawa.

C. BROOKE WORTH is one of the most remarkable men in the world of biological sciences. He is a doctor of medicine and a noted specialist on tropical viruses. He is an entomologist and one of the world's leading authorities on mosquitoes. His contributions on such widely diverse subjects as ornithology, parasitology, and mammalogy regularly appear in technical journals. And yet he writes with wit, wisdom, and wonder about all facets of the natural world. A graduate of the University of Pennsylvania School of Medicine, Brooke Worth served on the field staff of the Rockefeller Foundation in Florida, India, South Africa, and Trinidad, investigating insect-borne viruses, and from his experiences came two fascinating books, *A Naturalist in Trinidad* and *Mosquito Safari*. His most recent book, *Of Mosquitoes, Moths, and Mice*, recounts his adventures on his farm in New Jersey's Cape May County, with its rich population of biting bugs.